# Measurements and Analysis in
# Power System Harmonics

# Measurements and Analysis in Power System Harmonics

Edited by **Kim White**

CLANRYE
INTERNATIONAL

New Jersey

Published by Clanrye International,
55 Van Reypen Street,
Jersey City, NJ 07306, USA
www.clanryeinternational.com

**Measurements and Analysis in Power System Harmonics**
Edited by Kim White

International Standard Book Number: 978-1-63240-348-3 (Hardback)

Printed in the United States of America.

# Contents

# Preface

The concepts of measurements and analysis in power system harmonics are discussed in this elaborative book. The rise in the usage of power electronics equipment has led to significant distortions. An ideal AC power system is purely a sinusoidal wave, both voltage and current. However, the increasing existence of non-linear loads changes the aspects of voltage and current from the perfect sinusoidal wave. This alteration from the perfect wave is reflected by the harmonics and although its effects vary based upon the type of load, it affects the effectiveness of an electrical system and can create extensive damage to the system and infrastructure. Establishing ideal power quality after a good design and devices leads to productivity, efficiency, competitiveness and profitability. However, nobody can promise the ideal power quality when there is a good design, if the accurate tests and working methods from the collected information are not properly assured at every moment; this entails processing the actual information accurately. This book discusses harmonics analysis, ranging from the actual measurement data to the study of various industrial environments and electronic tools.

This book is a result of research of several months to collate the most relevant data in the field.

When I was approached with the idea of this book and the proposal to edit it, I was overwhelmed. It gave me an opportunity to reach out to all those who share a common interest with me in this field. I had 3 main parameters for editing this text:

1. Accuracy – The data and information provided in this book should be up-to-date and valuable to the readers.
2. Structure – The data must be presented in a structured format for easy understanding and better grasping of the readers.
3. Universal Approach – This book not only targets students but also experts and innovators in the field, thus my aim was to present topics which are of use to all.

Thus, it took me a couple of months to finish the editing of this book.

I would like to make a special mention of my publisher who considered me worthy of this opportunity and also supported me throughout the editing process. I would also like to thank the editing team at the back-end who extended their help whenever required.

**Editor**

# Part 1

## Measurements

# Electric Power Systems Harmonics - Identification and Measurements

Soliman Abdelhady Soliman[1]
and Ahmad Mohammad Alkandari[2]
*[1]Misr University for Science and Technology,*
*[2]College of Technological Studies,*
*[1]Egypt*
*[2]Kuwait*

## 1. Introduction

The presence of non-linear loads and the increasing number of distributed generation power systems (DGPS) in electrical grids contribute to change the characteristics of voltage and current waveforms in power systems, which differ from pure sinusoidal constant amplitude signals. Under these conditions advanced signal processing techniques are required for accurate measurement of electrical power quantities. The impact of non-linear loads in electrical power systems has been increasing during the last decades. Such electrical loads, which introduce non-sinusoidal current consumption patterns (current harmonics), can be found in rectification front-ends in motor drives, electronic ballasts for discharge lamps, personal computers or electrical appliances. Harmonics in power systems mean the existence of signals, superimposed on the fundamental signal, whose frequencies are integer numbers of the fundamental frequency. The electric utility companies should supply their customers with a supply having a constant frequency equal to the fundamental frequency, 50/60 Hz, and having a constant magnitude. The presence of harmonics in the voltage or current waveform leads to a distorted signal for voltage or current, and the signal becomes non-sinusoidal signal which it should not be. Thus the study of power system harmonics is an important subject for power engineers.

The power system harmonics problem is not a new problem; it has been noticed since the establishment of the ac generators, where distorted voltage and current waveforms were observed in the thirtieth of 20th century [2].

Concern for waveform distortion should be shared by all electrical engineers in order to establish the right balance between exercising control by distortion and keeping distortion under control. There is a need for early co-ordination of decisions between the interested parties, in order to achieve acceptable economical solutions and should be discussed between manufacturers, power supply and communication authorities [1].

Electricity supply authorities normally abrogate responsibility on harmonic matters by introducing standards or recommendations for the limitation of voltage harmonic levels at the points of common coupling between consumers.

## 2. Sources and problems of harmonics [2]

Harmonic sources are divided into two categories:
1.   Established and known
2.   New and Future

A review of the literature indicates that the known sources of harmonics include:
1.   Tooth ripple or ripples in the voltage waveform of rotating machines.
2.   Variations in air-gap reluctance over synchronous machine pole pitch.
3.   Flux distortion in the synchronous machine from sudden load changes.
4.   Non-sinusoidal distribution of the flux in the air gap of synchronous machines.
5.   Transformer magnetizing currents.
6.   Network nonlinearities from loads such as rectifiers, inverters, welders, arc furnaces, voltage controllers, frequency converters, etc.

While the established sources of harmonics are still present on the system, the power network is also subjected to new harmonic sources:
1.   Energy conservation measures, such as those for improved motor efficiency and load matching, which employ power semiconductor devices and switching for their operation. These devices often produce irregular voltage and current waveforms that are rich in harmonics.
2.   Motor control devices such as speed controls for traction.
3.   High-voltage direct-current power conversion and transmission.
4.   Interconnection of wind and solar power converters with distribution systems.
5.   Static var compensators which have largely replaced synchronous condensors as continuously variable-var sources.
6.   The development and potentially wide use of electric vehicles that require a significant amount of power rectification for battery charging.
7.   The potential use of direct energy conversion devices, such as magneto-hydrodynamics, storage batteries, and fuel cells that require dc/ac power converters.
8.   Cyclo-converters used for low-speed high-torque machines.
9.   Pulse-burst-modulated heating elements for large furnaces.

Today's power system harmonic problems can be traced to a number of factors:
1.   The substantial increase of nonlinear loads resulting from new technologies such as silicon-controlled rectifiers (SCRs), power transistors, and microprocessor controls which create load-generated harmonics throughout the system.
2.   A change in equipment design philosophy. In the past, equipment designs tended to be under-rated or over-designed. Now, in order to be competitive, power devices and equipment are more critically designed and, in the case of iron-core devices, their operating points are more into nonlinear regions. Operation in these regions results in a sharp rise in harmonics.

The most damaging frequencies to power devices and machines appear to be the lower – below 5-kHz – frequency range. In years past, the magnitudes and sources of these lower-frequency harmonics were limited and, inmost cases, power systems could tolerate them. The increase in power loss due to harmonics was also neglected because energy costs were low. These conditions no longer apply, and concern for harmonics is now becoming widespread among utilities.

For more Than 100 years, harmonics have been reported to cause operational problems to the power systems. Some of the major effects include:

1.  Capacitor bank failure from dielectric breakdown or reactive power overload.
2.  Interference with ripple control and power line carrier systems, causing mis-operation of systems which accomplish remote switching, load control, and metering.
3.  Excessive losses in – and heating of – induction and synchronous machines.
4.  Over voltages and excessive currents on the system from resonance to harmonic voltages or currents on the network.
5.  Dielectric breakdown of insulated cables resulting from harmonic over voltages on the system.
6.  Inductive interference with telecommunications systems.
7.  Errors in induction kWh meters.
8.  Signal interference and relay malfunction, particularly in solid-state and microprocessor-controlled systems.
9.  Interference with large motor controllers and power plant excitation systems. (Reported to cause motor problems as well as non-uniform output.)
10. Mechanical oscillations of induction and synchronous machines.
11. Unstable operation of firing circuits based on zero voltage crossing detection or latching.

These effects depend, of course, on the harmonic source, its location on the power system, and the network characteristics that promote propagation of harmonics.

## 3. Estimation of harmonics and sub-harmonics; the static case

### 3.1 Time domain model [3]

In this model, it is assumed that the waveform under consideration consists of a fundamental frequency component and harmonic components with order of integral multiples of the fundamental frequency. It is also assumed that the frequency is known and constant during the estimation period. Consider a non-sinusoidal voltage given by a Fourier-type equation:

$$v(t) = \sum_{n=0}^{N} V_n \sin(n\omega_0 t + \phi_n) \tag{1}$$

where

$v(t)$     is the instantaneous voltage at time $t$ (s.)
$V_n$     is the voltage amplitude of harmonic $n$
$\phi_n$     is the phase angle of harmonic $n$
$\omega_0$     is the fundamental frequency
$n$     order of harmonic
$N$     total number of harmonics

Equation (1) can be written as

$$v(t) = \sum_{n=0}^{N} (V_n \cos\phi_n \sin\omega_0 t + V_n \sin\phi_n \cos\omega_0 t) \tag{2}$$

Define

$$x_n = V_n \cos\phi_n \tag{3a}$$

$$y_n = V_n \sin \phi_n \tag{3b}$$

Then, equation (2) can be written as

$$v(t) = \sum_{n=0}^{N} \left( x_n \sin n\omega_0 t + y_n \cos n\omega_0 t \right) \tag{4}$$

If the voltage signal $v(t)$ is sampled at a pre-selected rate, say $\Delta t$, then $m$ samples would be obtained at $t_1$, $t_2 = t_1 + \Delta t$, $t_3 = t_1 + 2\Delta t$, ..., $t_m = t_1 + (m-1)\Delta t$. Then, after expanding equation (4), it can be written as

$$\begin{bmatrix} v(t_1) \\ v(t_2) \\ \vdots \\ v(t_m) \end{bmatrix} = \begin{bmatrix} a_{11}(t_1) & a_{12}(t_1) & \cdots & a_{12N+1}(t_1) \\ a_{21}(t_2) & a_{22}(t_2) & \cdots & a_{22N+1}(t_2) \\ \vdots & & & \\ a_{m1}(t_m) & a_{m2}(t_m) & \cdots & a_{m2N+1}(t_m) \end{bmatrix} \begin{bmatrix} y_0 \\ x_1 \\ y_1 \\ \vdots \\ y_N \end{bmatrix} \tag{5}$$

where the elements of the $A$ matrix are the sine and cosine expansion of equation (4). In the a's vector form, equation (5) can be written as

$$\underline{Z}_v(t) = A(t)\underline{\theta}_v + \underline{\epsilon}_v(t) \tag{6}$$

where $\underline{Z}_v(t)$ is $m \times 1$ vector of sampled voltage measurement, $A(t)$ is $m \times (2N + 1)$ matrix of measurement coefficients, $\underline{\theta}_v$ is $(2N + 1)$ vector to be estimated, $\epsilon_v(t)$ is $m \times 1$ error vector to be minimized. The order of the matrix $A(t)$ depends n the number of harmonics to be estimated. Furthermore, the elements of the matrix $A(t)$ depend on the initial sampling time $t_1$ the sampling rate $\Delta t$ and the data window size used in the estimation process. The matrix $A(t)$ can be calculated on off-line and stored.

At least $(2N + 1)$ samples are required to solve the problem formulated in (6). Using $2N + 1$ samples may produce a poor estimate, since we force $\epsilon_v(t)$ to be zero. We assume that $m > 2N + 1$, so that equation (4) represents over determined set of equations.

### 3.1.1 Time domain estimation; least error squares estimation (LES)

The solution to the over determined set of equations of (6) in the LES sense is given by

$$\begin{aligned} \underline{\theta}_v^* &= \left[ A^T(t)A(t) \right]^{-1} A^T(t)\underline{Z}_v(t) \\ &= A^+(t)\underline{Z}_v(t) \end{aligned} \tag{7}$$

where $A^+(t) = [A^T(t)A(t)]^{-1}A^T(t)$ is the left pseudo inverse. Having obtained the $\underline{\theta}_v^*$, the magnitude of any harmonic of order $n$ can be calculated as

$$V_n = \left[ x_n^2 + n_n^2 \right]^{\frac{1}{2}} ; n = 1, \ldots, N \tag{8}$$

while the phase angle of the $n$th harmonic is:

$$\phi_n = \tan^{-1} \frac{y_n}{x_n} \tag{9}$$

The above estimation procedures are simple and straight forward if the voltage and/or current waveforms under investigation are stationary and in steady state, but if there is a sudden variation in the power system operation, transient operation, such as fault, lighting and sudden loading to the system or sudden switching off a large load, the voltage signal waveform may contain, for a few cycles, a dc component, which if it is neglected, will affect the harmonics estimation content in the waveform. To overcome this problem, the voltage signal in equation (1) may be remodeled to take into account the dc component as [4]

$$v(t) = V_0 e^{-t/\tau} + \sum_{n=1}^{N} V_n \sin(n\omega_0 t + \phi) \tag{10}$$

where
$V_0$      is the amplitude of decaying dc component at $t = 0$
$\tau$      Is the time constant of the decaying dc component
The exponential term in equation (10) can be expanded using Taylor series and its first two terms can be used as

$$v(t) = V_0 - \left(\frac{V_0}{\tau}\right)t + \sum_{n=1}^{N} V_n \sin(n\omega_0 t + \phi_n) \tag{11}$$

Define the new parameters

$$x_{11} = V_0 \tag{12a}$$

$$x_{12} = \frac{V_0}{\tau} \tag{12b}$$

Then, equation (11) can be written as

$$v(t) = x_{11} - tx_{12} + \sum_{n=1}^{N} (x_n \sin n\omega_0 t + y_n \cos n\omega_0 t) \tag{13}$$

If the voltage $v(t)$ is samples at a pre-selected rate $\Delta t$, then $m$ sample would be obtained at $t_1$, $t_2 = t_1 + \Delta t$, ..., $t_m = 1 + (m - 1)\Delta t$, in this case equation (13) becomes

$$\underline{Z}(t) = B(t)\underline{Y} + \underline{\xi}(t) \tag{14}$$

where
$\underline{Z}(t)$      is the $m \times 1$ voltage samples
$\underline{B}(t)$      is $m \times (2N +2)$ measurement matrix whose elements depend on the initial and sampling times  and its order depends on the number of harmonics and the number of terms chosen from Taylor series expansion for the exponential term.
$\underline{Y}$      is $(2N +2) \times 1$ parameters vector to be estimated containing $x_{11}$, $x_{12}$ and $x_n$, $y_n$,
$\xi(t)$      is $m \times 1$ error vector to be minimized.
If $m > (2N +2)$, we obtain over determined set of equation and the non-recursive least error square algorithm can be used to solve this system of equation as

$$\underline{Y}^* = \left[B^T(t)B(t)\right]^{-1} B^T(t)\underline{Z}(t) \tag{15}$$

Having obtained the parameters vector $\underline{Y}^*$, the harmonics magnitude and phase angle can be obtained as

$$V_n = \left[ x_n^2 + y_n^2 \right]^{\frac{1}{2}} \tag{16}$$

$$\phi_n = \tan^{-1} \frac{y_n}{x_n} \tag{17}$$

while the parameters of the dc component can be calculated as

$$V_0 = x_{11} \tag{18a}$$

$$\tau = \frac{x_{11}}{x_{12}} \tag{18b}$$

Figure 1 gives actual recorded data for a three-phase dynamic load. The load is a variable frequency drive controlling a 3000 HP induction motor connected to an oil pipeline compressor [5]. Examining this curve reveals the following: (a) the waveform of the phase currents are not periodical; (b) there are low-frequency transients, which have frequencies not an integer number of the fundamental, we call them sub-harmonics, contaminating these waveforms, especially in the tips of the wave; and (c) the phase currents are not symmetrical. It can be concluded from these remarks that this waveform is contaminated with harmonics, as well as low frequency transients, this is due to the power electronic devices associated with the load.

## 3.2 Modeling of sub-harmonics in time domain
The sub-harmonics is a noise contaminated with a signal and having frequency which is not a multiple from the fundamental frequency (50/60 Hz), as given in equation (19). To measure these sub-harmonics, an accurate model is needed to present the voltage and current waves:

Assume the voltage or current waveform is contaminated with both harmonics and sub-harmonics. Then, the waveform can be written as

$$f(t) = \left[ A_1 e^{\sigma_1 t} \cos w_1 t + \sum_{i=2}^{N} A_i e^{\sigma_i t} \cos\left( w_i t + \varphi_i \right) \right] + \left[ \sum_{k=1}^{M} B_k \cos\left( w_k t + \theta_k \right) \right] \tag{19}$$

where

| | |
|---|---|
| $A_1, A_2, ..., A_N$ | are the sub-harmonics magnitude |
| $B_1, B_2, ..., B_k$ | are the harmonics magnitude |
| $\sigma_1, \sigma_2, ..., \sigma_N$ | are the damping constants |
| $\phi_i; i = 1, ..., N$ | are the sub-harmonic phase angles |
| $\theta_k; k = 1, ..., M$ | are the harmonic phase angles |
| $w_i; i=1, ..., N$ | are the sub-harmonic frequencies, assumed to be identified in the frequency domain |
| $w_k; k = 1, ..., M$ | are the harmonic frequencies assumed to be identified also in the frequency domain. |

Note that $w_i \neq w_k$, but $w_i = \left( \dfrac{w_1}{i} \right)$, $i = 3, ..., N$.

The first bracket in Equation (19) presents the possible low or high frequency sinusoidal with a combination of exponential terms, while the second bracket presents the harmonics, whose frequencies, $w_k$, $k = 1, ..., M$, are greater than 50/60 c/s, that contaminated the voltage or current waveforms. If these harmonics are identified to a certain degree of accuracy, i.e. a large number of harmonics are chosen, and then the first bracket presents the error in the voltage or current waveforms. Now, assume that these harmonics are identified, then the error $e(t)$ can be written as

$$e(t) = A_1 e^{\sigma_1 t} \cos w_1 t + \sum_{i=2}^{N} A_i e^{\sigma_i t} \cos \left( w_i t + \phi_i \right) \tag{20}$$

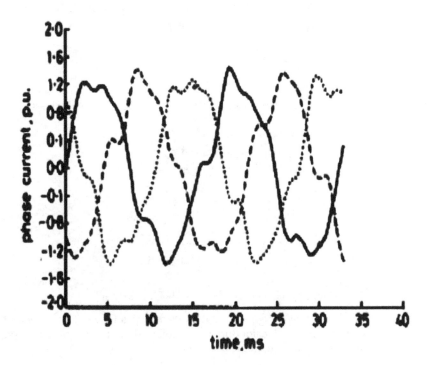

Fig. 1. Actual recorded phase currents.

It is clear that this expression represents the general possible low or high frequency dynamic oscillations. This model represents the dynamic oscillations in the system in cases such as, the currents of an induction motor when controlled by variable speed drive. As a special case, if the sampling constants are equal to zero then the considered wave is just a summation of low frequency components. Without loss of generality and for simplicity, it can be assumed that only two modes of equation (21) are considered, then the error $e(t)$ can be written as (21)

$$e(t) = A_1 e^{\sigma_1 t} \cos(w_1 t) + A_2 e^{\sigma_2 t} \cos(w_2 t + \phi_2) \tag{21}$$

Using the well-known trigonometric identity

$$\cos(w_2 t + \phi_2) = \cos w_2 t \cos \phi_2 - \sin w_2 t \sin \phi_2$$

then equation (21) can be rewritten as:

$$e(t) = A_1 e^{\sigma_1 t} \cos w_1 t + \left(e^{\sigma_2 t} \cos w_2 t\right) A_2 \cos \phi_2 + \left(e^{\sigma_2 t} \sin w_2 t\right) A_2 \sin \phi_2 \tag{22}$$

It is obvious that equation (22) is a nonlinear function of $A$'s, $\sigma$'s and $\phi$'s. By using the first two terms in the Taylor series expansion $A_i e^{\sigma_i t}$; $i = 1,2$. Equation (22) turns out to be

$$e(t) = A_1 \cos w_1 t + (t \cos w_1 t)(A_1 \sigma_1) + (\cos w_2 t)(A_2 \cos \phi_2) + (t \cos w_2 t)(A_2 \sigma_2 \cos \phi_2)$$
$$- (\sin w_2 t)(A_2 \sin \phi_2) + (t \sin w_2 t)(A_2 \sigma_2 \sin \phi_2) \tag{23}$$

where the Taylor series expansion is given by:

$$e^{\sigma t} = 1 + \sigma t$$

Making the following substitutions in equation (23), equation (26) can be obtained,

$$\begin{cases} x_1 = A_1; & x_2 = A_1 \sigma_1 \\ x_3 = A_2 \cos \phi_2; & x_4 = A_2 \sigma_2 \cos \phi_2 \\ x_5 = A_2 \sin \phi_2; & x_6 = A_2 \sigma_2 \sin \phi_2 \end{cases} \tag{24}$$

and

$$\begin{cases} h_{11}(t) = \cos w_1 t; & h_{12}(t) = t \cos w_1 t \\ h_{13}(t) = \cos w_2 t; & h_{14}(t) = t \cos w_2 t \\ h_{15}(t) = -\sin w_2 t; & h_{16}(t) = -t \sin w_2 t \end{cases} \tag{25}$$

$$e(t) = h_{11}(t) x_1 + h_{12}(t) x_2 + h_{13}(t) x_3 + h_{14}(t) x_3 + h_{15}(t) x_4 + h_{16}(t) x_5 \tag{26}$$

If the function $f(t)$ is sampled at a pre-selected rate, its samples would be obtained at equal time intervals, say $\Delta t$ seconds. Considering m samples, then there will be a set of $m$ equations with an arbitrary time reference $t_1$ given by

$$\begin{vmatrix} e(t_1) \\ e(t_2) \\ \dots \\ \dots \\ e(t_m) \end{vmatrix} = \begin{vmatrix} h_{11}(t_1) & h_{12}(t_1) & \dots & h_{16}(t_1) \\ h_{21}(t_2) & h_{22}(t_2) & \dots & h_{26}(t_2) \\ \dots & \dots & \dots & \dots \\ \dots & \dots & \dots & \dots \\ h_{m1}(t_m) & h_{m2}(t_m) & \dots & h_{m6}(t_m) \end{vmatrix} \begin{vmatrix} x_1 \\ x_2 \\ \dots \\ \dots \\ x_6 \end{vmatrix} \tag{27}$$

It is clear that this set of equations is similar to the set of equations given by equation (5). Thus an equation similar to (6) can be written as:

$$\underline{z}(t) = H(t)\underline{\theta}(t) + \xi(t) \tag{28}$$

where $z(t)$ is the vector of sampled measurements, $H(t)$ is an $m \times 6$, in this simple case, matrix of measurement coefficients, $\theta(t)$ is a $6 \times 1$ parameter vector to be estimated, and $\xi(t)$ is an $m \times 1$ noise vector to be minimized. The dimensions of the previous matrices depend on the number of modes considered, as well as, the number of terms truncated from the Taylor series.

### 3.2.1 Least error squares estimation
The solution to equation (28) based on LES is given as

$$\underline{\theta}^*(t) = \left[H^T(t)H(t)\right]^{-1} H^T(t)\underline{Z}(t) \tag{29}$$

Having obtained the parameters vector $\underline{\theta}^*(t)$, then the sub-harmonics parameters can be obtained as

$$A_1 = x_1^*, \quad \sigma_1 = \frac{x_2^*}{x_1^*} \tag{30}$$

$$A_2 = \left[x_3^{*2} + x_5^{*2}\right]^{\frac{1}{2}}, \quad \sigma_2 = \frac{x_4^*}{x_3^*} \tag{31}$$

$$\tan\phi_2 = \frac{x_5^*}{x_3^*} = \frac{x_6^*}{x_4^*} \tag{32}$$

### 3.2.2 Recursive least error squares estimates
In the least error squares estimates explained in the previous section, the estimated parameters, in the three cases, take the form of

$$\left[\underline{\theta}^*\right]_{n\times1}^{m-1} = [A]_{n\times m}^+ [Z]_{m\times1} \tag{33}$$

where $[A]^+$ is the left pseudo inverse of $[A] = [A^TA]^{-1}A^T$, the superscript "$m - 1$" in the equation represents the estimates calculated using data taken from $t = t_1$ to $t = t_1 + (m - 1)\Delta t$ s, $t_1$ is the initial sampling time. The elements of the matrix $[A]$ are functions of the time reference, initial sampling time, and the sampling rate used. Since these are selected in advance, the left pseudo inverse of $[A]$ can be determined for an application off-line.

Equation 33 represents, as we said earlier, a non recursive least error squares (LES) filter that uses a data window of m samples to provide an estimate of the unknowns, $\theta$. The estimates of $[\theta]$ are calculated by taking the row products of the matrix $[A]^+$ with the $m$ samples. A new sample is included in the data window at each sampling interval and the oldest sample is discarded. The new $[A]^+$ for the latest m samples is calculated and the estimates of $[\theta]$ are updated by taking the row products of the updated $[A]^+$ with the latest m samples. However, equation (33) can be modified to a recursive form which is computationally more efficient.

Recall that equation

$$[Z]_{m\times1} = [A]_{m\times n}[\theta]_{n\times1} \tag{34}$$

represents a set of equations in which $[Z]$ is a vector of m current samples taken at intervals of $\Delta t$ seconds. The elements of the matrix $[A]$ are known. At time $t = t_1 + m\Delta t$ a new sample is taken. Then equation (33) can be written as

$$\left[\theta^*\right]_{n\times1}^m = \left[\frac{A}{a_{mi}}\right]_{n\times(mH)}^+ \left[\frac{[Z]}{[Z_m]}\right]_{(mH)\times1} \tag{35}$$

where the superscript "$m$" represents the new estimate at time $t = t_1 + m\Delta t$. It is possible to express the new estimates obtained from equation (34) in terms of older estimates (obtained from equation (33)) and the latest sample $Z_m$ as follows

$$\left[\theta^*\right]^m = \left[\theta^*\right]^{m-1} + \left[\alpha(m)\right]\left[\left[Z_m\right]-\left[a_{mi}\right]\left[\theta^*\right]^{m-1}\right] \tag{36a}$$

This equation represents a recursive least squares filter. The estimates of the vector $[\theta]$ at $t = t_1 + m\Delta t$ are expressed as a function of the estimates at $t = t_1 + (m-1)\Delta t$ and the term $\left[\left[Z_m\right]-\left[a_{mi}\right]\left[\theta^*\right]^{m-1}\right]$. The elements of the vector, $[\alpha(m)]$, are the time-invariant gains of the recursive least squares filter and are given as

$$\alpha(m) = \left[A^T A\right]^{-1}\left[a_{mi}\right]^T\left[[I]+\left[a_{mi}\right]\left[A^T A\right]^{-1}\left[a_{mi}\right]^T\right]^{-1} \tag{36b}$$

### 3.2.3 Least absolute value estimates (LAV) algorithm (Soliman & Christensen algorithm) [3]

The LAV estimation algorithm can be used to estimate the parameters vectors. For the reader's convenience, we explain here the steps behind this algorithm.

Given the observation equation in the form of that given in (28) as

$$\underline{Z}(t) = A(t)\underline{\theta} + \in(t)$$

The steps in this algorithm are:

**Step 1.** Calculate the LES solution given by

$$\left[\theta^*\right] = \left[A(t)\right]^+\underline{Z}(t), \left[A(t)\right]^+ = \left[A^T(t)A(t)\right]^{-1}A^T(t)$$

**Step 2.** Calculate the LES residuals vector generated from this solution as

$$\underline{r}^* = \underline{Z}(t) - A(t)\left[A(t)\right]^+ \underline{Z}(t)$$

**Step 3.** Calculated the standard deviation of this residual vector as

$$\sigma = \frac{1}{m-n+1}\left[\sum_{i=1}^{m}(r_i - \bar{r})^2\right]^{\frac{1}{2}}$$

Where $\bar{r} = \frac{1}{m}\sum_{i=1}^{m} r_i$ , the average residual

**Step 4.** Reject the measurements having residuals greater than the standard deviation, and recalculate the LES solution

**Step 5.** Recalculated the least error squares residuals generated from this new solution

**Step 6.** Rank the residual and select $n$ measurements corresponding to the smallest residuals

**Step 7.** Solve for the LAV estimates $\hat{\theta}$ as

$$\left[\hat{\theta}\right]^*_{n\times1} = \left[\hat{A}(t)\right]^{-1}_{n\times n}\hat{Z}_{n\times1}(t)$$

**Step 8.** Calculate the LAV residual generated from this solution

### 3.3 Computer simulated tests

Ref. 6 carried out a comparative study for power system harmonic estimation. Three algorithms are used in this study; LES, LAV, and discrete Fourier transform (DFT). The data used in this study are real data from a three-phase six pulse converter. The three techniques are thoroughly analyzed and compared in terms of standard deviation, number of samples and sampling frequency.

For the purpose of this study, the voltage signal is considered to contain up to the 13th harmonics. Higher order harmonics are neglected. The rms voltage components are given in Table 1.

| RMS voltage components corresponding to the harmonics | | | | | |
|---|---|---|---|---|---|
| Harmonic frequency | Fundamental | 5th | 7th | 11th | 13th |
| Voltage magnitude (p.u.) | 0.95∠−2.02° | 0.09∠82.° | 0.043∠8.9° | 0.030∠212.9° | 0.033∠162.6° |

Table 1.

Figure 2 shows the A.C. voltage waveform at the converter terminal. The degree of the distortion depends on the order of the harmonics considered as well as the system characteristics. Figure 3 shows the spectrum of the converter bus bar voltage.

The variables to be estimated are the magnitudes of each voltage harmonic from the fundamental to the 13th harmonic. The estimation is performed by the three techniques while several parameters are changed and varied. These parameters are the standard deviation of the noise, the number of samples, and the sampling frequency. A Gaussian-distributed noise of zero mean was used.

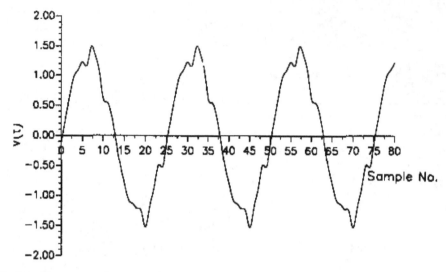

Fig. 2. AC voltage waveform

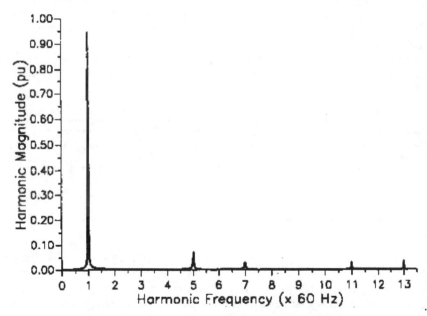

Fig. 3. Frequency spectrums.

Figure 4 shows the effects of number of samples on the fundamental component magnitude using the three techniques at a sampling frequency = 1620 Hz and the measurement set is corrupted with a noise having standard deviation of 0.1 Gaussian distribution.

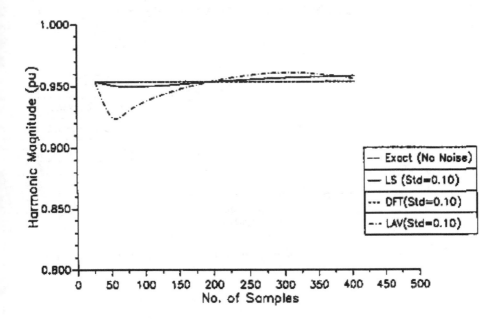

Fig. 4. Effect of number of samples on the magnitude estimation of the fundamental harmonic (sampling frequency = 1620 Hz).

It can be noticed from this figure that the DFT algorithm gives an essentially exact estimate of the fundamental voltage magnitude. The LAV algorithm requires a minimum number of samples to give a good estimate, while the LES gives reasonable estimates over a wide range of numbers of samples. However, the performance of the LAV and LES algorithms is improved when the sampling frequency is increased to 1800 Hz as shown in Figure 5.

Figure 6 –9 gives the same estimates at the same conditions for the 5th, 7th, 11th and 13th harmonic magnitudes. Examining these figures reveals the following remarks.

- For all harmonics components, the DFT gives bad estimates for the magnitudes. This bad estimate is attributed to the phenomenon known as "spectral leakage" and is due to the fact that the number of samples per number of cycles is not an integer.
- As the number of samples increases, the LES method gives a relatively good performance. The LAV method gives better estimates for most of the number of samples.
- At a low number of samples, the LES produces poor estimates.

However, as the sampling frequency increased to 1800 Hz, no appreciable effects have changed, and the estimates of the harmonics magnitude are still the same for the three techniques.

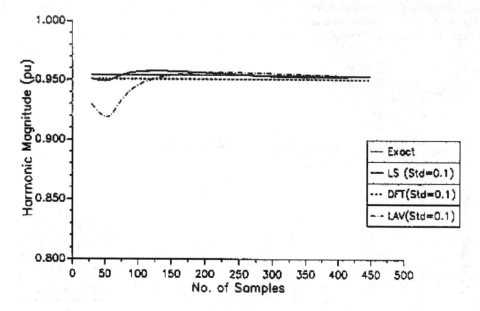

Fig. 5. Effect of number of samples on the magnitude estimation of the fundamental harmonic (sampling frequency = 1800 Hz).

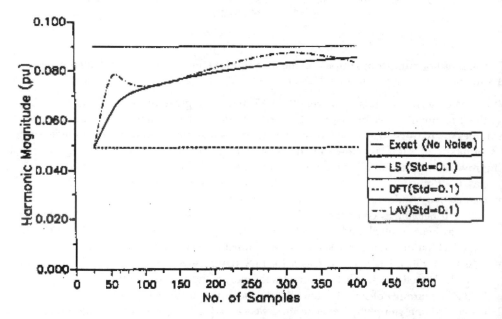

Fig. 6. Effect of number of samples on the magnitude estimation of the 5th harmonic (sampling frequency = 1620 Hz).

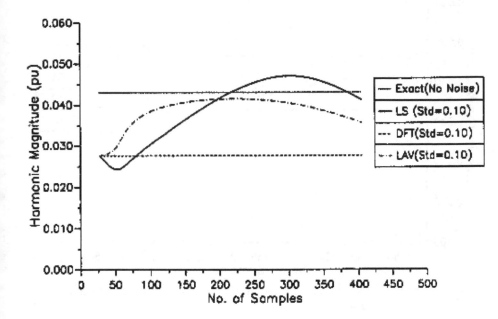

Fig. 7. Effect of number of samples on the magnitude estimation of the 7th harmonic (sampling frequency = 1620 Hz).

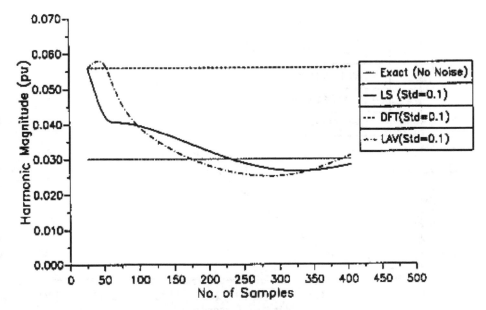

Fig. 8. Effect of number of samples on the magnitude estimation of the 11th harmonic (sampling frequency = 1620 Hz).

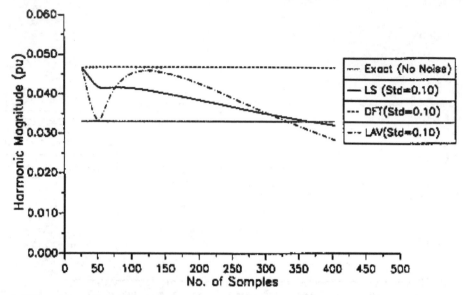

Fig. 9. Effect of number of samples on the magnitude estimation of the 13th harmonic (sampling frequency = 1620 Hz).

The CPU time is computed for each of the three algorithms, at a sampling frequency of 1620 Hz. Figure 10 gives the variation of CPU.

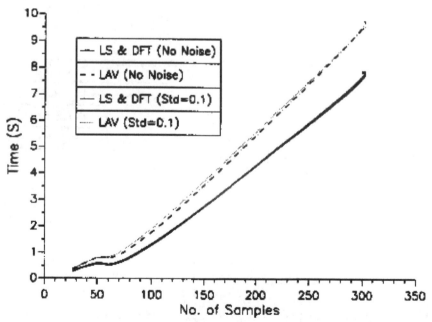

Fig. 10. The CPU times of the LS, DFT, and LAV methods (sampling frequency = 1620 Hz).

The CPU time for the DFT and LES algorithms are essentially the same, and that of the LAV algorithm is larger. As the number of samples increases, the difference in CPU time between the LAV and LS/DFT algorithm increases.

Other interesting studies have been carried out on the performance of the three algorithms when 10% of the data is missed, taken uniformly at equal intervals starting from the first data point, for the noise free signal and 0.1 standard deviation added white noise Gaussian, and the sampling frequency used is 1620 Hz.

Figure 11 gives the estimates of the three algorithms at the two cases. Examining this figure we can notice the following remarks:

For the no noise estimates, the LS and DFT produce bad estimates for the fundamental harmonic magnitude, even at a higher number of samples

The LAV algorithm produces good estimates, at large number of samples.

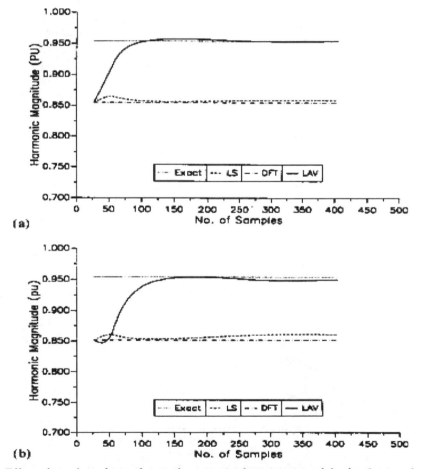

Fig. 11. Effect of number of samples on the magnitude estimation of the fundamental harmonic for 10% missing data (sampling frequency = 1620 Hz): (a) no noise; (b) 0.1 standard deviation added white Gaussian noise.

Figure 12 –15 give the three algorithms estimates, for 10% missing data with no noise and with 0.1 standard deviation Gaussian white noise, when the sampling frequency is 1620 Hz for the harmonics magnitudes and the same discussions hold true.

## 3.4 Remarks

Three signal estimation algorithms were used to estimate the harmonic components of the AC voltage of a three-phase six-pulse AC-DC converter. The algorithms are the LS, LAV, and DFT. The simulation of the ideal noise-free case data revealed that all three methods give exact estimates of all the harmonics for a sufficiently high sampling rate. For the noisy case, the results are completely different. In general, the LS method worked well for a high number of samples. The DFT failed completely. The LAV gives better estimates for a large range of samples and is clearly superior for the case of missing data.

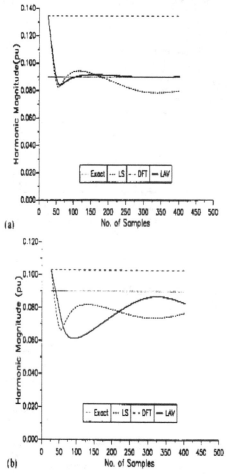

Fig. 12. Effect of number of samples on the magnitude estimation of the 5th harmonic for 10% missing data (sampling frequency = 1620 Hz): (a) no noise; (b) 0.1 standard deviation added white Gaussian noise.

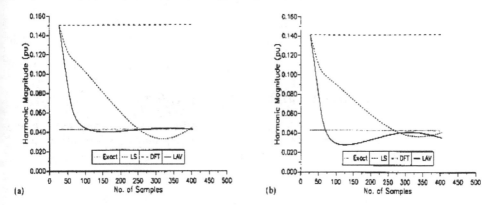

(a)    (b)

Fig. 13. Effect of number of samples on the magnitude estimation of the 7th harmonic for 10% missing data (sampling frequency = 1620 Hz): (a) no noise; (b) 0.1 standard deviation added white Gaussian noise.

## 4. Estimation of harmonics; the dynamic case

In the previous section static-state estimation algorithms are implemented for identifying and measuring power system harmonics. The techniques used in that section was the least error squares (LES), least absolute value (LAV) and the recursive least error squares algorithms. These techniques assume that harmonic magnitudes are constant during the data window size used in the estimation process. In real time, due to the switching on-off of power electronic equipments (devices) used in electric derives and power system transmission (AC/DC transmission), the situation is different, where the harmonic magnitudes are not stationary during the data window size. As such a dynamic state estimation technique is required to identifying (tracking) the harmonic magnitudes as well as the phase angles of each harmonics component.

In this section, we introduce the Kalman filtering algorithm as well as the dynamic least absolute value algorithm (DLAV) for identifying (tracking) the power systems harmonics and sub-harmonics (inter-harmonics).

The Kalman filtering approach provides a mean for optimally estimating phasors and the ability to track-time-varying parameters.

The state variable representation of a signal that includes n harmonics for a noise-free current or voltage signal $s(t)$ may be represented by [7]

$$s(t) = \sum_{i=1}^{n} A_i(t) \cos(i\omega t + \theta_i) \tag{37}$$

where

$A_i(t)$     is the amplitude of the phasor quantity representing the $i$th harmonic at time t

$\theta_i$     is the phase angle of the $i$th harmonic relative to a reference rotating at $i\omega$

$n$     is the harmonic order

Each frequency component requires two state variables. Thus the total number of state variable is 2n. These state variables are defined as follows

$$x_1(t) = A_1(t)\cos\theta_1, \qquad x_2(t) = A_1(t)\sin\theta_1$$
$$x_2(t) = A_2(t)\cos\theta_2, \qquad x_3(t) = A2_1(t)\sin\theta_2$$
$$\cdots \qquad\qquad\qquad \cdots \qquad\qquad , \qquad (38)$$
$$\cdots \qquad\qquad\qquad \cdots$$
$$x_{2n-1}(t) = A_n(t)\cos\theta_n, \quad x_{2n}(t) = A_2(t)\sin\theta_n$$

These state variables represent the in-phase and quadrate phase components of the harmonics with respect to a rotting reference, respectively. This may be referred to as model 1. Thus, the state variable equations may be expressed as:

$$\begin{bmatrix} x_1 \\ x_2 \\ \vdots \\ x_{2n-1} \\ x_{2n} \end{bmatrix} = \begin{bmatrix} 1 & 0 & \cdots & \cdots & 0 \\ 0 & 1 & \cdots & \cdots & 0 \\ \vdots & \vdots & \vdots & \vdots & \vdots \\ 0 & 0 & \cdots & 1 & 0 \\ 0 & 0 & \cdots & \cdots & 1 \end{bmatrix} \begin{bmatrix} x_1 \\ x_2 \\ \vdots \\ x_{2n-1} \\ x_{2n} \end{bmatrix} + \begin{bmatrix} \alpha_1 \\ \alpha_2 \\ \vdots \\ \alpha_{2n-1} \\ \alpha_{2n} \end{bmatrix} w_k \qquad (39)$$

or in short hand

$$\underline{X}(k+1) = \phi\underline{X}(k) + \underline{w}(k) \qquad (40)$$

where
$\underline{X}$        is a $2n \times 1$ state vector
$\phi$        Is a $2n \times 2n$ state identity transition matrix, which is a diagonal matrix
$\underline{w}(k)$       is a $2n \times 1$ noise vector associated with the transition of a sate from $k$ to $k+1$ instant
The measurement equation for the voltage or current signal, in this case, can be rewritten as, equation (37)

$$s(k\Delta t) = \begin{bmatrix} \cos wk\Delta t & \sin wk\Delta t & \cdots & \cos nwk\Delta t & \sin(nwk\Delta t) \end{bmatrix} \begin{bmatrix} x_1 \\ x_2 \\ \vdots \\ x_{2n-1} \\ x_{2n} \end{bmatrix}_k + v(k) \qquad (41)$$

which can be written as

$$\underline{Z}(k) = H(k)\underline{X}(k) + \underline{v}(k) \qquad (42)$$

where $\underline{Z}(\underline{k})$ is $m \times 1$ vector of measurements of the voltage or current waveforms, $H(k)$ is $m \times 2n$ measurement matrix, which is a time varying matrix and $\underline{v}(k)$ is $m \times 1$ errors measurement vector. Equation (40) and (42) are now suitable for Kalman filter application. Another model can be derived of a signal with time-varying magnitude by using a stationary reference, model 2. Consider the noise free signal to be

$$s(t_k) = A(t_k)\cos(wt + \theta) \qquad (43)$$

Now, consider $x_1(k) = A(t_k)\cos(wt_k + \theta)$ and $x_2(k)$ to be $A(t_k)\sin(wt_k + \theta)$. At $t_{k+2}$, which is $t_k + \Delta t$, the signal may be expressed as

$$s(t_{k+1}) = A(t_{k+1})\cos(wt_k + w\Delta t + \theta) = x_1(k+1)$$
$$x_1(k+1) = x_1(k)\cos(w\Delta t) - x_2(k)\sin(w\Delta t)$$

also

$$x_2(k+1) = A(t_{k+1})\sin(wt_k + w\Delta t + \theta)$$
$$= x_1(k)\sin(w\Delta t) + x_2(k)\cos(w\Delta t)$$

Thus, the state variable representation takes the following form

$$\begin{bmatrix} x_1(k+1) \\ x_2(k+1) \end{bmatrix} = \begin{bmatrix} \cos w\Delta t & -\sin w\Delta t \\ \sin w\Delta t & \cos w\Delta t \end{bmatrix} \begin{bmatrix} x_1(k) \\ x_2(k) \end{bmatrix} + \begin{bmatrix} w_1(k) \\ w_2(k) \end{bmatrix} \tag{44}$$

and the measurement equation then becomes

$$Z(k) = \begin{bmatrix} 1 & 0 \end{bmatrix} \begin{bmatrix} x_1(k) \\ x_2(k) \end{bmatrix} + v(k) \tag{45}$$

If the signal includes $n$ frequencies; the fundamental plus $n - 1$ harmonics, the state variable representation may be expressed as

$$\begin{bmatrix} x_1(k+1) \\ x_2(k+1) \\ \vdots \\ x_{2n-1}(k+1) \\ x_{2n}(k+1) \end{bmatrix} = \begin{bmatrix} M_1 & \cdots & \cdots & 0 \\ \cdots & \cdots & \cdots & \cdots \\ \cdots & \cdots & \cdots & \cdots \\ 0 & \cdots & \cdots & M_n \end{bmatrix} \begin{bmatrix} x_1(k+1) \\ x_2(k+1) \\ \vdots \\ x_{2n-1}(k+1) \\ x_{2n}(k+1) \end{bmatrix} + \begin{bmatrix} \alpha_1 \\ \alpha_2 \\ \vdots \\ \alpha_{2n-1} \\ \alpha_{2n} \end{bmatrix} w(k) \tag{46}$$

where the sub-matrices $M_i$ are given as

$$M_i = \begin{bmatrix} \cos iw\Delta t & -\sin iw\Delta t \\ \sin iw\Delta t & \cos iw\Delta t \end{bmatrix}, i = 1,\ldots,n \tag{47}$$

Equation (46) can be rewritten as

$$\underline{X}(k+1) = \phi(k)\underline{X}(k) + \underline{w}(k) \tag{48}$$

while equation (45) as

$$\underline{Z}(k) = H\underline{X}(k) + \underline{V}(k) \tag{49}$$

This model has constant state transition and measurement matrices. However, it assumes a stationary reference. Thus, the in-phase and quadrature phase components represent the instantaneous values of con-sinusoidal and sinusoidal waveforms, respectively.

## 4.1 Testing the kalman filter algorithm

The two Kalman filter models described in the preceding section were tested using a waveform with known harmonic contents. The waveform consists of the fundamental, the third, the fifth, the ninth, the eleventh, the thirteenth, and the nineteenth harmonics. The waveform is described as

$$s(t) = 1.0\cos\left(\omega t + 10°\right) + 0.1\cos\left(3\omega t + 20°\right) + 0.08\cos\left(5\omega t + 30°\right)$$
$$+0.08\cos\left(7\omega t + 40°\right) + 0.06\cos\left(11\omega t + 50°\right)$$
$$+0.05\cos\left(13\omega t + 60°\right) + 0.03\cos\left(19\omega t + 70°\right)$$

The sampling frequency was selected to be 64 × 60 Hz.

i.   Initial process vector

As the Kalman filter model started with no past measurement, the initial process vector was selected to be zero. Thus, the first half cycle (8 milliseconds) is considered to be the initialization period.

ii.  Initial covariance matrix

The initial covariance matrix was selected to be a diagonal matrix with the diagonal values equal to 10 p.u.

iii. Noise variance ($R$)

The noise variance was selected to be constant at a value of 0.05 p.u.$^2$. This was passed on the background noise variance at field measurement.

iv.  State variable covariance matrix ($Q$)

The matrix $Q$ was also selected to be 0.05 p.u.

Testing results of model 1, which is a 14-state model described by equations (40) and (42) are given in the following figures. Figure 14 shows the initialization period and the recursive estimation of the magnitude of the fundamental and third harmonic. Figure 15 shows the Kalman gain for the fundamental component. Figure 16 shows the first and second diagonal element of $P_k$.

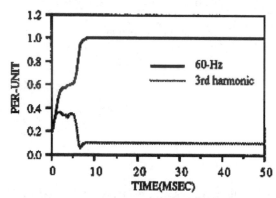

Fig. 14. Estimated magnitudes of 60 Hz and third harmonic component using the 14-state model 1.

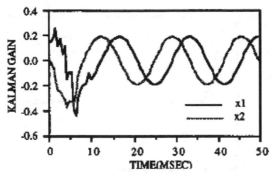

Fig. 15. Kalman gain for $x_1$ and $x_2$ using the 14-state model 1.

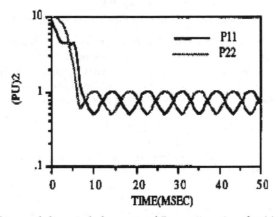

Fig. 16. The first and second diagonal elements of $P_k$ matrix using the 14-state model 1.

While the testing results of model 2 are given in Figures 26 –28. Figure 26 shows the first two components of Kalman gain vector. Figure 27 shows the first and second diagonal elements of $P_k$. The estimation of the magnitude of and third harmonic were exactly the same as those shown in Figure 23.

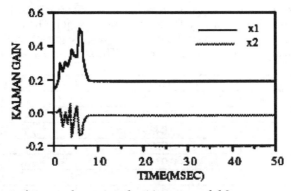

Fig. 17. Kalman gain for $x_1$ and $x_2$ using the 14-state model 2.

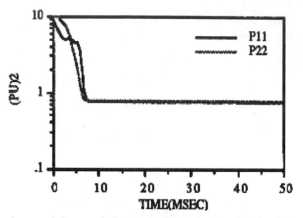

Fig. 18. The first and second diagonal elements of $P_k$ matrix using the 14-state model 2.

The Kalman gain vector $K_k$ and the covariance matrix $P_k$ reach steady-state in about half a cycle, when model 1 is used, 1/60 seconds. Its variations include harmonics of 60 Hz. The covariance matrix in the steady-state consists of a constant plus a periodic component. These time variations are due to the time-varying vector in the measurement equation. Thus, after initialization of the model, the Kalman gain vector of the third cycle can be repeated for successive cycles.

When model 2 is selected, the components of the Kalman gain vector and the covariance matrix become constants. In both models, the Kalman gain vector is independent of the measurements and can be computed off-line. As the state transition matrix is a full matrix, it requires more computation than model 1 to update the state vector.

Kalman filter algorithm is also tested for actual recorded data. Two cases of actual recorded data are reported here. The first case represents a large industrial load served by two parallel transformers totaling 7500 KVA [5]. The load consists of four production lines of induction heating with two single-phase furnaces per line. The induction furnaces operate at 8500 Hz and are used to heat 40-ft steel rods which are cut into railroad spikes. Diodes are used in the rectifier for converting the 60 Hz power into dc and SCRs are used in the inverter for converting the dc into single-phase 8500 Hz power. The waveforms were originally sampled at 20 kHz. A program was written to use a reduced sampling rate in the analysis. A careful examination of the current and voltage waveforms indicated that the waveforms consist of (1) harmonics of 60 Hz and (2) a decaying periodic high-frequency transients. The high-frequency transients were measured independently for another purpose [6]. The rest of the waveform was then analyzed for harmonic analysis. Using a sampling frequency that is a multiple of 2 kHz, the DFT was then applied for a period of 3 cycles. The DFT results were as follows:

| Freq. (Hz) | Mag. | Angle (rad.) |
|---|---|---|
| 60 | 1.0495 | -0.20 |
| 300 | 0.1999 | 1.99 |
| 420 | 0.0489 | -2.18 |
| 660 | 0.0299 | 0.48 |
| 780 | 0.0373 | 2.98 |
| 1020 | 0.0078 | -0.78 |
| 1140 | 0.0175 | 1.88 |

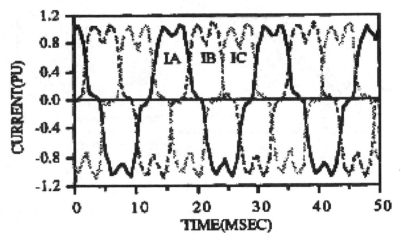

Fig. 19. Actual recorded current waveform of phases A, B, and C.

The Kalman filter, however, can be applied for any number of samples over a half cycle. If the harmonic has time-varying magnitude, the Kalman filter algorithm would track the time variation after the initialization period (half a cycle). Figures 19 and 20 show the three-phase current and voltage waveforms recorded at the industrial load. Figures 21 –23 show the recursive estimation of the magnitude of the fundamental, fifth, and seventh harmonics; the eleventh and thirteenth harmonics; and the seventeenth and nineteenth harmonics, respectively, for phase A current. The same harmonic analysis was also applied to the actual recorded voltage waveforms. Figure 24 shows the recursive estimation of the magnitude of the fundamental and fifth harmonic for phase A voltage. No other voltage harmonics are shown here due tot he negligible small value.

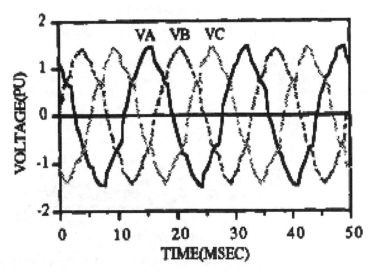

Fig. 20. Actual recorded voltage waveform of phase A, B, and C.

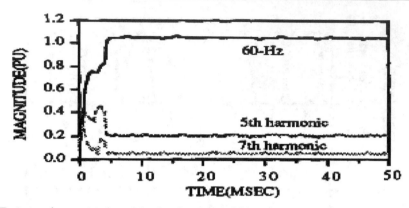

Fig. 21. Estimated magnitudes of the fundamental, fifth, and seventh harmonics for phase A current.

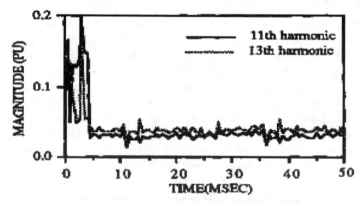

Fig. 22. Estimated magnitudes of the eleventh and thirteenth harmonics for phase A current.

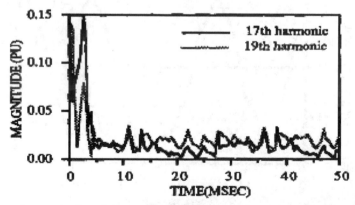

Fig. 23. Estimated magnitudes of the seventeenth and nineteenth harmonics for phase A current.

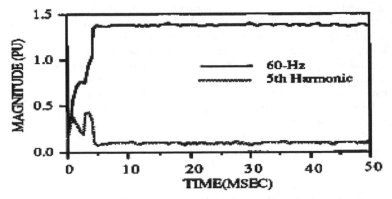

Fig. 24. Estimated magnitudes of the 60 Hz and fifth harmonic for phase A voltage.

The second case represents a continuous dynamic load. The load consists of two six-phase drives for two 200 HP dc motors. The current waveform of one phase is shown in Figure 25. The harmonic analysis using the Kalman filter algorithm is shown in Figure 35. It should be noted that the current waveform was continuously varying in magnitude due to the dynamic nature of the load. Thus, the magnitude of the fundamental and harmonics were continuously varying. The total harmonic distortion experienced similar variation.

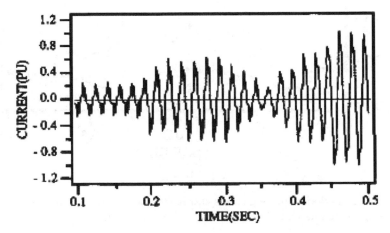

Fig. 25. Current waveform of a continuous varying load.

There is no doubt that the Kalman filtering algorithm is more accurate and is not sensitive to a certain sampling frequency. As the Kalman filter gain vector is time0varying, the estimator can track harmonics with the time varying magnitudes.

Two models are described in this section to show the flexibility in the Kalman filtering scheme. There are many applications, where the results of FFT algorithms are as accurate as a Kalman filter model. However, there are other applications where a Kalman filter becomes superior to other algorithms. Implementing linear Kalman filter models is relatively a simple task. However, state equations, measurement equations, and covariance matrices need to be correctly defined.

Kalman filter used in the previous section assumes that the digital samples for the voltage and current signal waveforms are known in advance, or at least, when it is applied on-line, good estimates for the signals parameters are assumed with a certain degree of accuracy, so that the filter converges to the optimal estimates in few samples later. Also, it assumes that an accurate model is presented for the signals; otherwise inaccurate estimates would be obtained. Ref. 8 uses the Kalman filter algorithm to obtain the optimal estimate of the power system harmonic content. The measurements used in this reference are the power system voltage and line flows at different harmonics obtained from a harmonic load flow program (HARMFLO). The effect of load variation over a one day cycle on the power system harmonics and standard are presented. The optimal estimates, in this reference, are the power system bus voltage magnitudes and phase angles at different harmonic level.

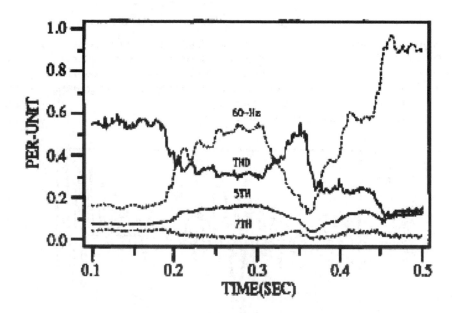

Fig. 35. Magnitude of dominant frequencies and harmonic distortion of waveform shown in Figure 34 using the Kalman filtering approach.

## 4.2 Linear dynamic weighted least absolute estimates [11]

This section presents the application of the linear dynamic weighted least absolute value dynamic filter for power system harmonics identification and measurements. The two models developed earlier, model 1 and model 2, are used with this filter. As we explained earlier, this filter can deal easily with the outlier, unusual events, in the voltage or current waveforms.

*Software implementation*

A software package has been developed to analyze digitized current and voltage waveforms. This package has been tested on simulated data sets, as well as on an actual recorded data set. and computes the voltage and current harmonics magnitude, the voltage and current harmonics phase angles, and the fundamental power and harmonics power.

*Initialization of the filter*

To initialize the recursive process of the proposed filter, with an initial process vector and covariance matrix $P$, a simple deterministic procedure uses the static least squares error estimate of previous measurements. Thus, the initial process vector may be computed as:

$$\hat{X}_0 = \left[H^T H\right]^{-1} H^T z$$

and the corresponding covariance error matrix is:

$$\hat{P}_0 = \left[H^T H\right]^{-1}$$

where $H$ is an $m \times m$ matrix of measurements, and $z$ is an $m \times 1$ vector of previous measurements, the initial process vector may be selected to be zero, and the first few milliseconds are considered to be the initialization period.

## 4.3 Testing the algorithm using simulated data

The proposed algorithm and the two models were tested using a voltage signal waveform of known harmonic contents described as:

$$v(t) = 1\cos(\omega t + 10°) + 0.1\cos(3\omega t + 20°) + 0.08\cos(5\omega t + 30°) + 0.08\cos(9\omega t + 40°)$$
$$+0.06\cos(11\omega t + 50°) + 0.05\cos(13\omega t + 60°) + 0.03\cos(19\omega t + 70°)$$

The data window size is two cycles, with sampling frequency of 64 samples/cycle. That is, the total number of samples used is 128 samples, and the sampling frequency is 3840 Hz. For this simulated example we have the following results.

Using the two models, the proposed filtering algorithm estimates exactly the harmonic content of the voltage waveform both magnitudes and phase angles and the two proposed models produce the same results.

The steady-state gain of the proposed filter is periodic with a period of 1/60 s. This time variation is due to the time varying nature of the vector states in the measurement equation. Figure 54 give the proposed filter gain for $X_1$ and $Y_1$.

The gain of the proposed filter reaches the steady-state value in a very short time, since the initialization of the recursive process, as explained in the preceding section, was sufficiently accurate.

The effects of frequency drift on the estimate are also considered. We assume small and large values for the frequency drift: $\Delta f$ = -0.10 Hz and $\Delta f$ = -1.0 Hz, respectively. In this study the elements of the matrix $H(k)$ are calculated at 60 Hz, and the voltage signal is sampled at ($\omega = 2\pi f$, $f$ = 60 + $\Delta f$). Figs. 24 and 29 give the results obtained for these two frequency deviations for the fundamental and the third harmonic. Fig. 55 gives the estimated magnitude, and Fig. 29 gives the estimated phase angles. Examination of these two curves reveals the following:

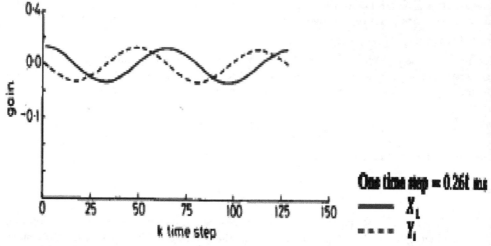

Fig. 27. Gain of the proposed filter for $X_1$ and $Y_1$ using models 1 and 2.

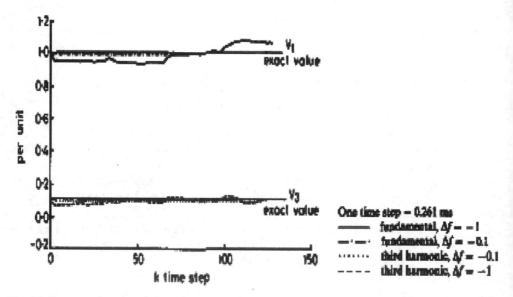

Fig. 28. Estimated magnitudes of 60 Hz and third harmonic for frequency drifts using models 1 and 2.

- For a small frequency drift, $\Delta f$ = -0.10 Hz, the fundamental magnitude and the third harmonic magnitude do not change appreciably; whereas for a large frequency drift, $\Delta f$ = -1.0 Hz, they exhibit large relative errors, ranging from 7% for the fundamental to 25% for the third harmonics.
- On the other hand, for the small frequency drift the fundamental phase angle and the third harmonic phase angle do not change appreciably, whereas for the large frequency

drift both phase angles have large changes and the estimates produced are of bad quality.

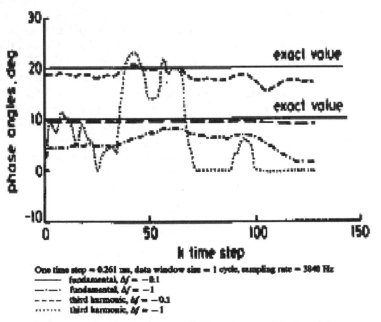

Fig. 29. Estimated phase angles for frequency drifts using models 1 and 2

To overcome this drawback, it has been found through extensive runs that if the elements of the matrix $H(k)$ are calculated at the same frequency of the voltage signal waveform, good estimates are produced and the frequency drift has in this case no effect. Indeed, to perform this modification the proposed algorithm needs a frequency-measurement algorithm before the estimation process is begun.

It has been found, through extensive runs that the filter gains for the fundamental voltage components, as a case study, do not change with the frequency drifts. Indeed, that is true since the filter gain $K(k)$ does not depend on the measurements (eqn. 8).

As the state transition matrix for model 2 is a full matrix, it requires more computation than model 1 to update the state vector. Therefore in the rest of this study, only model 1 is used.

### 4.4 Testing on actual recorded data

The proposed algorithm is implemented to identify and measure the harmonics content for a practical system of operation. The system under study consists of a variable-frequency drive that controls a 3000 HP, 23 kV induction motor connected to an oil pipeline compressor. The waveforms of the three phase currents are given in Fig. 31. It has been found for this system that the waveforms of the phase voltages are nearly pure sinusoidal waveforms. A careful examination of the current waveforms revealed that the waveforms consist of: harmonics of 60 Hz, decaying period high-frequency transients, and harmonics of less than 60 Hz (sub-harmonics). The waveform was originally sampled at a 118 ms time interval and a sampling frequency of 8.5 kHz. A computer program was written to change this sampling rate in the analysis.

Figs. 31 and 32 show the recursive estimation of the magnitude of the fundamental, second, third and fourth harmonics for the voltage of phase $A$. Examination of these curves reveals that the highest-energy harmonic is the fundamental, 60 Hz, and the magnitude of the second, third and fourth harmonics are very small. However, Fig. 33 shows the recursive estimation of the fundamental, and Fig. 34 shows the recursive estimation of the second, fourth and sixth harmonics for the current of phase A at different data window sizes. Indeed, we can note that the magnitudes of the harmonics are time-varying since their magnitudes change from one data window to another, and the highest energy harmonics are the fourth and sixth. On the other hand, Fig. 35 shows the estimate of the phase angles of the second, fourth and sixth harmonics, at different data window sizes. It can be noted from this figure that the phase angles are also time0varing because their magnitudes vary from one data window to another.

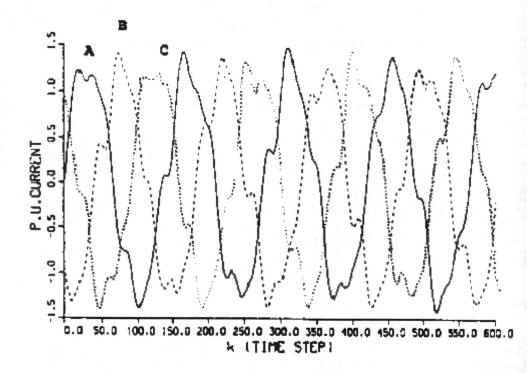

Fig. 30. Actual recorded current waveform of phases $A$, $B$ and $C$.

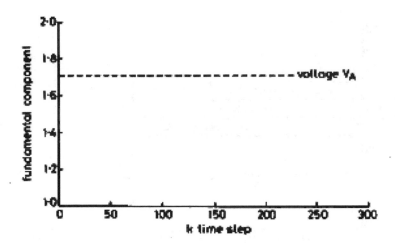

Fig. 31. Estimated fundamental voltage.

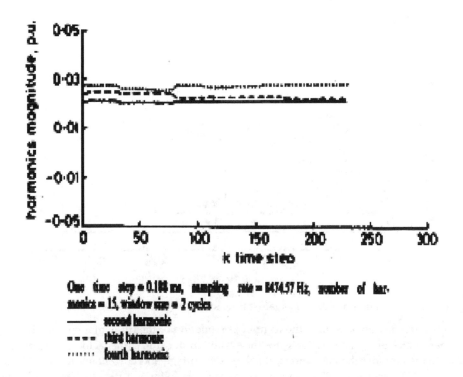

Fig. 32. Estimated voltage harmonics for $V$

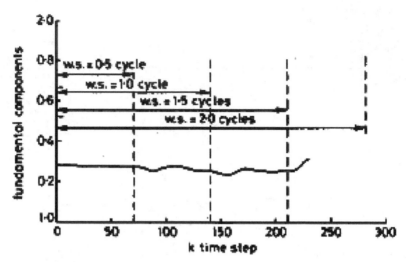

Fig. 33. Estimated fundamental current $I_A$.

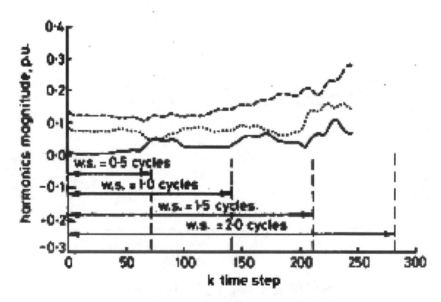

Fig. 34. Harmonics magnitude of $I_A$ against time steps at various window sizes.

Furthermore, Figs. 36 – 38 show the recursive estimation of the fundamental, fourth and the sixth harmonics power, respectively, for the system under study (the factor 2 in these figures is due to the fact that the maximum values for the voltage and current are used to calculate this power). Examination of these curves reveals the following results. The fundamental power and the fourth and sixth harmonics are time-varying.

For this system the highest-energy harmonic component is the fundamental power, the power due to the fundamental voltage and current.

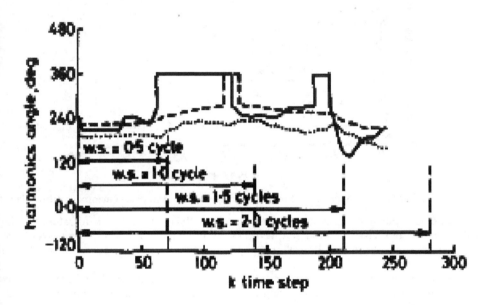

Fig. 35. Harmonics phase angles of $I_A$ against time steps at various window sizes.

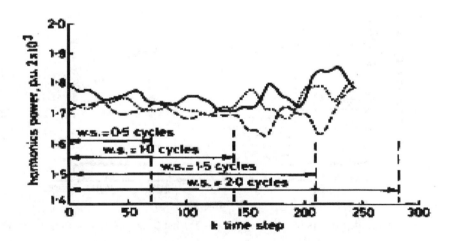

Fig. 36. Fundamental powers against time steps.

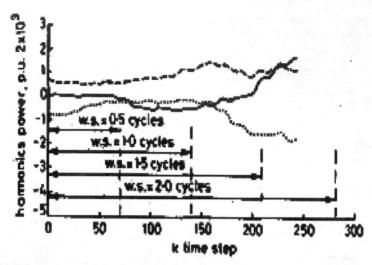

Fig. 37. Fourth harmonic power in the three phases against time steps at various window sizes.

The fundamental powers, in the three phases, are unequal; i.e. the system is unbalanced. The fourth harmonic of phase C, and later after 1.5 cycles of phase A, are absorbing power from the supply, whereas those for phase B and the earlier phase A are supplying power to the network.

The sixth harmonic of phase B is absorbing power from the network, whereas the six harmonics of phases A and C are supplying power to the network; but the total power is still the sum of the three-phase power.

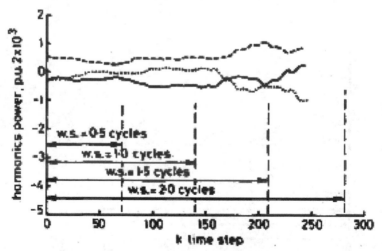

Fig. 38. Sixth harmonic powers in the three phases against time steps at various window sizes.

The fundamental power and the fourth and sixth harmonics power are changing from one data window to another.

## 4.5 Comparison with Kalman Filter (KF) algorithm

The proposed algorithm is compared with KF algorithm. Fig 39 gives the results obtained when both filters are implemented to estimate the second harmonic components of the current in phase *A*, at different data window sizes and when the considered number of harmonics is 15. Examination of the Figure reveals the following; both filters produce almost the same estimate for the second harmonic magnitude; and the magnitude of the estimated harmonic varies from one data window to another.

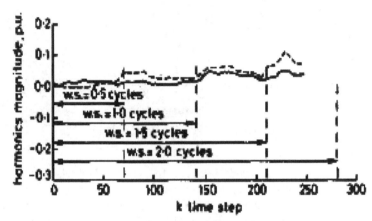

Fig. 39. Estimated second harmonic magnitude using KF and WLAV.

### 4.5.1 Effects of outliers

In this Section the effects of outliers (unusual events on the system waveforms) are studied, and we compare the new proposed filter and the well-known Kalman filtering algorithm. In the first Subsection we compare the results obtained using the simulated data set of Section 2, and in the second Subsection the actual recorded data set is used.

### Simulated data

The simulated data set of Section 4.3 has been used in this Section, where we assume (randomly) that the data set is contaminated with gross error, we change the sign for some measurements or we put these measurements equal to zero. Fig. 40 shows the recursive estimate of the fundamental voltage magnitude using the proposed filter and the well-known Kalman filtering algorithm. Careful examination of this curve reveals the following results.

The proposed dynamic filter and the Kalman filter produce an optimal estimate to the fundamental voltage magnitude, depending on the data considered. In other words, the voltage waveform magnitude in the presence of outliers is considered as a time-varying magnitude instead of a constant magnitude.

The proposed filter and the Kalman filter take approximately two cycles to reach the exact value of the fundamental voltage magnitude. However, if such outliers are corrected, the discrete least absolute value dynamic filter almost produces the exact value of the fundamental voltage during the recursive process, and the effects of the outliers are greatly reduced Figure 41.

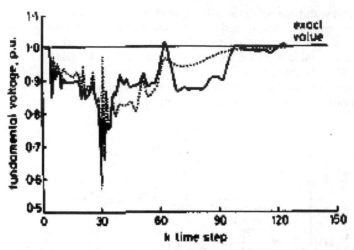

Fig. 40. Effects of bad data on the estimated fundamental voltage.

**Actual recorded data**

In this Section the actual recorded data set that is available is tested for outliers' contamination. Fig. 42 shows the recursive estimate of the fundamental current of phase A using the proposed filter, as well as Kalman filter algorithms. Indeed, both filters produce an optimal estimate according to the data available. However, if we compare this figure with Fig. 42, we can note that both filters produce an estimate different from what it should be. Fig. 42 shows the recursive estimates using both algorithms when the outliers are corrected. Indeed, the proposed filter produces an optimal estimate similar to what it should be, which is given in Fig. 43.

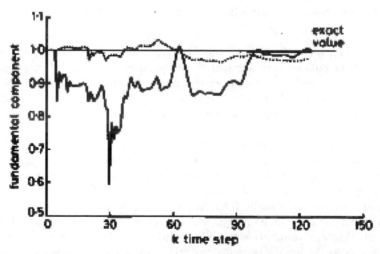

Fig. 41. Estimated fundamental voltage magnitude before and after correction for outliers.

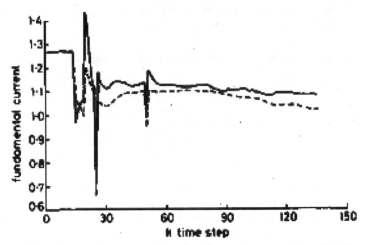

Fig. 42. Estimated fundamental current when the data set is contaminated with outliers.

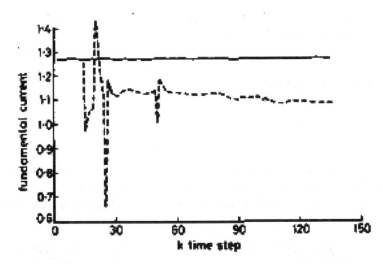

Fig. 43. Estimated fundamental current before and after correction for outliers.

### 4.6 Remarks

- The discrete least absolute dynamic filter (DLAV) can easily handle the parameters of the harmonics with time-varying magnitudes.
- The DLAV and KF produce the same estimates if the measurement set is not contaminated with bad data.
- The DLAV is able to identify and correct bad data, whereas the KF algorithm needs pre-filtering to identify and eliminate this bad data.

It has been shown that if the waveform is non-stationary, the estimated parameters are affected by the size of the data window.

It has been pointed out in the simulated results that the harmonic filter is sensitive to the deviations of frequency of the fundamental component. An algorithm to measure the power system frequency should precede the harmonics filter.

## 5. Power system sub-harmonics (interharmonics); dynamic case

As we said in the beginning of this chapter, the off-on switching of the power electronics equipment in power system control may produce damped transients of high and/or low frequency on the voltage and/or current waveforms. Equation (20) gives the model for such voltage waveform. The first term in this equation presents the damping inter-harmonics model, while the second term presents the harmonics that contaminated the voltage waveform including the fundamental. In this section, we explain the application of the linear dynamic Kalman filtering algorithm for measuring and identifying these inter-harmonics. As we said before, the identification process is split into two sub-problems. In the first problem, the harmonic contents of the waveform are identified. Once the harmonic contents of the waveform are identified, the reconstructed waveform can be obtained and the error in the waveform, which is the difference between the actual and the reconstructed waveform, can be obtained. In the second problem, this error is analyzed to identify the sub-harmonics.

Finally, the final error is obtained by subtracting the combination of the harmonic and the sub-harmonic contents, the total reconstructed, from the actual waveform. It has been shown that by identifying these sub-harmonics, the final error is reduced greatly.

### 5.1 Modeling of the system sub-harmonics

For Kalman filter application, equation (28) is the measurement equation, and we recall it here as

$$\underline{Z}(t) = \underline{H}(t)\underline{\theta}(t) + \underline{\omega}(t) \qquad (28)$$

If the voltage is sampled at a pre-selected rate, its samples would be obtained at equal time intervals, say $\Delta t$ seconds. Then equation (26) can be written at stage $k$, $k = 1, 2, ..., k$, where $K$ is the total number of intervals, $K$ = [window size in seconds/$\Delta t$] = [window size in seconds × sampling frequency (Hz)].

$$z(k\Delta t) = h_{11}(k\Delta t)x_1(k) + h_{12}(k\Delta t)x_2(k) + ... + h_{16}(k\Delta t)x_6(k) \qquad (50)$$

If there are m samples, equation (8.64) turns out to be a set of equations. Each equation defines the system at a certain time $(k\Delta t)$.

$$z_1(k\Delta t) = H_i(k\Delta t)\theta(k) + w_i(k) ; i = 1, 2, ..., m \qquad (51)$$

This equation can be written in vector form as:

$$z(k\Delta t) = H(k\Delta t)\theta(k) + w(k) \qquad (52)$$

where
$z(k)$        is $m \times 1$ measurement vector taken over the window size

$\theta(k)$       is $n \times 1$ state vector to be estimated. It could be harmonic or sub-harmonic parameters depending on both $H(k)$ and $z(k)$

$H(k)$       is $m \times n$ matrix giving the ideal connection between $z(k\Delta t)$ and $\theta(k)$ in the absence of noise $w(k)$. If the elements of $H(k\Delta t)$ are given by equation (25), it is clear that $H(k\Delta t)$ is a time-varying matrix.

$w(k)$       is an m noise vector to be minimized and is assumed to be random white noise with known covariance construction.

Equation (52) describes the measurement system equation at time $k\Delta t$.
The state space variable equation for this model may be expressed as $\Xi$

$$\begin{vmatrix} x_1(k+1) \\ x_2(k+1) \\ x_3(k+1) \\ ... \\ ... \\ x_u(k+1) \end{vmatrix} = \begin{vmatrix} x_1(k) \\ x_2(k) \\ x_3(k) \\ ... \\ ... \\ x_u(k) \end{vmatrix} + w(k) \tag{53}$$

Equation (67) can be rewritten in vector form as:

$$\theta(k+1) = \phi(k)\theta(k) + w(k) \tag{54}$$

where

$\phi(k)$       is $n \times n$ state transition matrix and it is an identity matrix

$w(k)$       is $n \times 1$ plant noise vector

Together equation (52) and (54) form the system dynamic model. It is worthwhile to state here that in this state space representation the time reference was chosen as a rotating time reference which caused the state transition matrix to be the identity matrix and the $H$ matrix to be a time varying matrix.

Having estimated the parameter vector $\theta$, the amplitude, damping constant, and the phase angle can be determined using equations (30) to (32), at any step

## 5.2 Testing kalman filter algorithm
### 5.2.1 Description of the load

The proposed algorithm is tested on an actual recorded data to obtain the damped sub-harmonics which contaminated the three phase current waveforms of a dynamic load. The load is a variable frequency drive controlling a 3000 HP induction motor connected to an oil pipe line compressor. The solid state drive is of 12 pulses designed with harmonic filter. The data given is the three phase currents at different motor speed, and is given in per unit. The three phase currents are given in Figure 42. This figure shows high harmonics in each phase current as well as sub-harmonics. It is clear that the currents have variable magnitudes from one cycle to another (non-stationary waveforms).

### 5.2.2 Sub-harmonic estimation

After the harmonic contents of the waveforms had been estimated, the waveform was reconstructed to get the error in this estimation. Figure 71 gives the real current and the reconstructed current for phase A as well as the error in this estimation. It has been found that the error has a maximum value of about 10%. The error signal is analyzed again to find if there are any sub-harmonics in this signal. The Kalman filtering algorithm is used here to find the amplitude and the phase angle of each sub-harmonic frequency. It was found that the signal has sub-harmonic frequencies of 15 and 30 Hz. The sub-harmonic amplitudes are given in Figure 43 while the phase angle of the 30 Hz component is given in Figure 44. The sub-harmonic magnitudes were found to be time varying, without any exponential decay, as seen clearly in Figure 43.

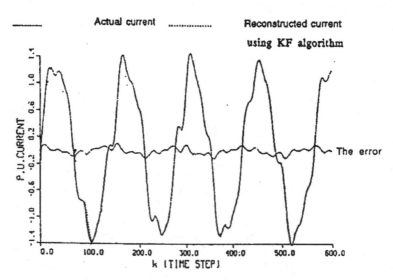

Fig. 42. Actual and reconstructed current for phase *A*

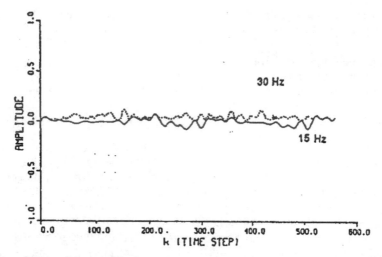

Fig. 43. The sub-harmonic amplitudes.

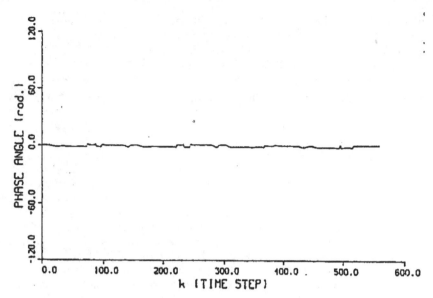

Fig. 44. The phase angle of the 30 Hz component.

Once the sub-harmonic parameters are estimated, the total reconstructed current can be obtained by adding the harmonic contents to the sub-harmonic contents. Figure 45 gives the total resultant error which now is very small, less than 3%.

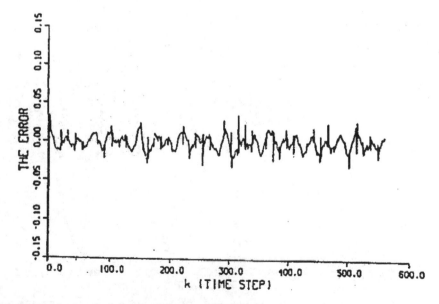

Fig. 45. the final error in the estimate using KF algorithm.

### 5.2.3 Remarks

- Kalman filter algorithm is implemented, in this section, for identification of sub-harmonics parameters that contaminated the power system signals.
- By identifying the harmonics and sub-harmonics of the signal under investigation, the total error in the reconstructed waveform is reduced greatly.

### 5.3 SUb-harmonics indentification with DLAV algorithm (Soliman and Christensen algorithm)

In this section, the application of discrete dynamic least absolute value algorithm for identification and measurement of sub-harmonics is discussed. The model used with Kalman filter algorithm and explained earlier, will be used in this section, a comparison with Kalman filter is offered at the end of this section. No needs to report here, the dynamic equations for the DLAV filter, since they are already given in the previous chapters, in the place, where we need them. The steps used with Kalman filter in the previous section, will be followed here. Hence, we discuss the testing of the algorithm.

As we said earlier, the algorithm first estimates the harmonics that contaminated one of the phases current waveforms, say phase $A$; in this estimation, we assumed a large number of harmonics. The reconstructed waveform and the error for this estimation, which is the difference between the actual recorded data and the reconstructed waveform, are then obtained. Figure 46 gives the real current and the reconstructed current for phase $A$, as well as the resultant error. The maximum error in this estimation was found to be about 13%. This error signal is then analyzed to identify the sub-harmonic parameters. Figure 47 gives the sub-harmonic amplitudes for sub-harmonic frequencies for 15 and 30 Hz, while Fig. 48 gives the phase angle estimate for the 30 Hz sub-harmonic. Note that in the sub-harmonic estimation process we assume that the frequencies of these sub-harmonics are known in advance, and hence the matrix $H$ can easily be formulated in an off-line mode.

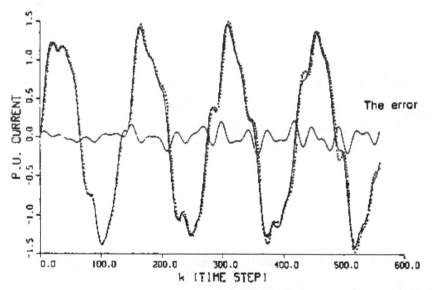

Fig. 46. Actual (full curve) and reconstructed (dotted curve) current for phase $A$ using the WLAVF algorithm.

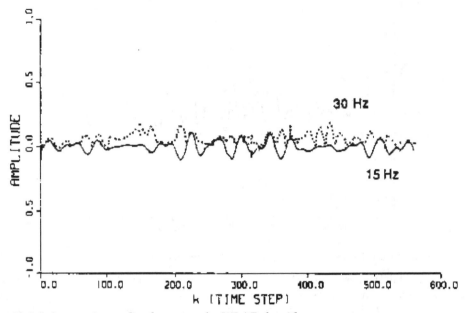

Fig. 47. Sub-harmonic amplitudes using the WLAF algorithm.

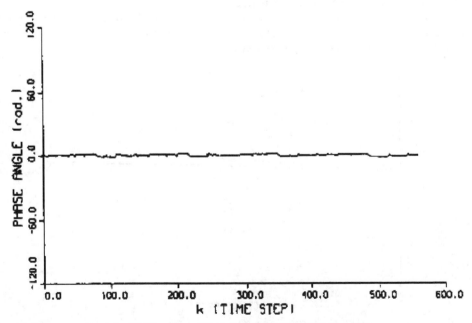

Fig. 48. Phase angle of the 30 Hz component using the WLAVF algorithm.

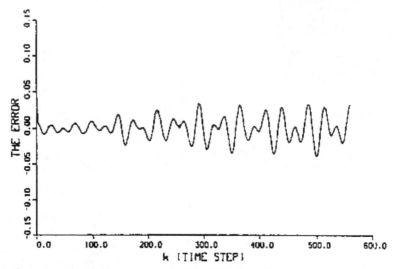

Fig. 49. Final error in the estimate using the WLAVF algorithm.

Finally, the total error is found by subtracting the combination of the harmonic and sub-harmonic contents, the total reconstructed waveform, from the actual waveform. This error is given in Fig. 49. It is clear from this Figure that the final error is very small, with a maximum value of about 3%.

## 5.4 Comparison between DLAV and KF algorithm

The proposed WLAVF algorithm was compared with the well-known linear KF algorithm. It can be shown that if there is no gross error contaminating the data, both filters produce very close results. However, some points may be mentioned here.

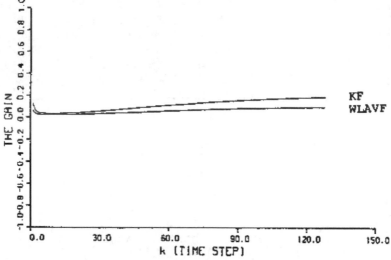

Fig. 50. Comparison between the filter gains for component $x_1$.

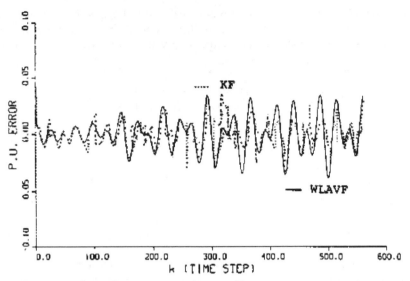

Fig. 51. Final errors in the estimation using the two filters.

1. The estimate obtained via the WLAVF algorithm is damped more than that obtained via the KF algorithm. This is probably due to the fact that the WLAVF gain is more damped and reaches a steady state faster than the KF gain, as shown in Fig. 50.
2. The overall error in the estimate was found to be very close in both cases, with a maximum value of about 3%. The overall error for both cases is given in Fig. 51.
3. Both algorithms were found to act similarly when the effects of the data window size, sampling frequency and the number of harmonics were studied

## 6. Park's transformation

Park's transformation is well known in the analysis of electric machines, where the three rotating phases abc are transferred to three equivalent stationary dq0 phases (d-q reference frame). This section presents the application of Park's transformation in identifying and measuring power system harmonics. The technique does not need a harmonics model, as well as number of harmonics expected to be in the voltage or current signal. The algorithm uses the digitized samples of the three phases of voltage or current to identify and measure the harmonics content in their signals. Sampling frequency is tied to the harmonic in question to verify the sampling theorem. The identification process is very simple and easy to apply.

### 6.1 Identification processes

In the following steps we assume that m samples of the three phase currents or voltage are available at the preselected sampling frequency that satisfying the sampling theorem. i.e. the sampling frequency will change according to the order of harmonic in question, for example if we like to identify the 9th harmonics in the signal. In this case the sampling frequency must be greater than 2*50*90=900 Hz and so on.

The forward transformation matrix at harmonic order n; n=1,2,.., N, N is the total expected harmonics in the signal, resulting from the multiplication of the modulating matrix to the αβ signal and the αβ0- transformation matrix is given as (dqo transformation or Park`s transformation)

$$P = \begin{bmatrix} \sin n\omega t & \cos n\omega t & 0 \\ \cos n\omega t & \sin n\omega t & 0 \\ 0 & 0 & 1 \end{bmatrix} \times$$

$$\sqrt{\frac{2}{3}} \begin{bmatrix} 1 & -0.5 & -0.5 \\ 0 & \dfrac{\sqrt{3}}{2} & \dfrac{-\sqrt{3}}{2} \\ \dfrac{1}{\sqrt{2}} & \dfrac{1}{\sqrt{2}} & \dfrac{1}{\sqrt{2}} \end{bmatrix} = \sqrt{\frac{2}{3}} \begin{bmatrix} \sin n\omega t & \sin(n\omega t + 120n) & \sin(n\omega t + 240n) \\ \cos n\omega t & \cos(n\omega t + 120n) & \cos(n\omega t + 240n) \\ \dfrac{1}{\sqrt{2}} & \dfrac{1}{\sqrt{2}} & \dfrac{1}{\sqrt{2}} \end{bmatrix} \qquad (55)$$

The matrix of equation (69) can be computed off line if the frequencies of the voltage or current signal as well as the order of harmonic to be identified are known in advance as well as the sampling frequency and the number of samples used. If this matrix is multiplied digitally by the samples of the three-phase voltage and current signals sampled at the same sampling frequency of matrix (55), a new set of three -phase samples are obtained, we call this set a dq0 set (reference frame). This set of new three phase samples contains the ac component of the three-phase voltage or current signals as well as the dc offset. The dc off set components can be calculated as;

$$V_d(dc) = \frac{1}{m} \sum_{i=1}^{m} (V_d)_i$$

$$V_q(dc) = \frac{1}{m} \sum_{i=1}^{m} (V_q)_i \qquad (56)$$

$$V_0(dc) = \frac{1}{m} \sum_{i=1}^{m} (V_o)_i$$

If these dc components are eliminated from the new pqo set, a new ac harmonic set is produced. We call this set as $V_d(ac)$, $V_q(ac)$ and $V_0(ac)$. If we multiply this set by the inverse of the matrix of equation (56), which is given as:

$$P^{-1} = \sqrt{\frac{2}{3}} \begin{bmatrix} \sin n\omega t & \cos n\omega t & \dfrac{1}{\sqrt{2}} \\ \sin(n\omega t + 240n) & \cos(n\omega t + 240n) & \dfrac{1}{\sqrt{2}} \\ \sin(n\omega t + 120n) & \cos(n\omega t + 120n) & \dfrac{1}{\sqrt{2}} \end{bmatrix} \qquad (57)$$

Then, the resulting samples represent the samples of the harmonic components in each phase of the three phases. The following are the identification steps.

1. Decide what the order of harmonic you would like to identify, and then adjust the sampling frequency to satisfy the sampling theory. Obtain m digital samples of harmonics polluted three-phase voltage or current samples, sampled at the specified sampling frequency $F_s$. Or you can obtain these m samples at one sampling frequency that satisfies the sampling theorem and cover the entire range of harmonic frequency you expect to be in the voltage or current signals. Simply choose the sampling frequency to be greater than double the highest frequency you expect in the signal

2. Calculate the matrices, given in equations (55) and (57) at m samples and the order of harmonics you identify. Here, we assume that the signal frequency is constant and equal the nominal frequency 50 or 60 Hz.

3. Multiplying the samples of the three-phase signal by the transformation matrix given by equation (57)

4. Remove the dc offset from the original samples; simply by subtracting the average of the new samples generated in step 2 using equation (56) from the original samples. The generated samples in this step are the samples of the ac samples of dqo signal.

5. Multiplying the resulting samples of step 3 by the inverse matrix given by equation (57). The resulting samples are the samples of harmonics that contaminate the three phase signals except for the fundamental components.

6. Subtract these samples from the original samples; we obtain m samples for the harmonic component in question

7. Use the least error squares algorithm explained in the preceding section to estimate the amplitude and phase angle of the component. If the harmonics are balanced in the three phases, the identified component will be the positive sequence for the 1st, 4th, 7th,etc and no negative or zero sequence components. Also, it will be the negative sequence for the 2nd, 5th , 8th etc component, and will be the zero sequence for the 3rd  6th, 9th etc components. But if the expected harmonics in the three phases are not balanced go to step 8.

8. Replace ω by -ω in the transformation matrix of equation (55) and the inverse transformation matrix of equation (57). Repeat steps 1 to 7 to obtain the negative sequence components.

## 6.2 Measurement of magnitude and phase angle of harmonic component

Assume that the harmonic component of the phase a voltage signal is presented as:

$$v_a(t) = V_{am} \cos(n\omega t + \phi_a) \tag{58}$$

where $V_{am}$ is the amplitude of  harmonic component n in phase a, ω is the fundamental frequency and $\phi_a$ its phase angle measured with respect to certain reference. Using the trigonometric identity, equation (58) can be written as:

$$v_a(t) = x_a \cos n\omega t + y_a \sin n\omega t \tag{59}$$

where we define

$$x_a = V_{am} \cos \phi_a \tag{60}$$

$$x_a = V_{am} \sin \phi_a \tag{61}$$

As stated earlier in step 5 m samples are available for a harmonic component of phase a, sampled at a preselected rate, then equation (73) can be written as:

$$Z = A\theta + \varepsilon \tag{62}$$

Where $Z$ is mx1 samples of the voltage of any of the three phases, A is mx2 matrix of measurement and can be calculated off line if the sampling frequencies as well as the signal frequency are known in advance. The elements of this matrix are;

$a_1(t) = \cos n\omega t, a_2(t) = \sin n\omega t$ ; $\theta$ is a 2x1 parameters vector to be estimated and $\zeta$ is mx1 error vector due to the filtering process to be minimized. The solution to equation (62) based on least error squares is

$$\theta^* = \left[ A^T A \right]^{-1} A^T Z \tag{63}$$

Having identified the parameters vector $\theta^*$ the magnitude and phase angle of the voltage of phase a can be calculated as follows:

$$V_{am} = \left[ x^2 + y^2 \right]^{\frac{1}{2}} \tag{64}$$

$$\phi_a = \tan^{-1} \frac{y}{x} \tag{65}$$

## 6.3 Testing the algorithm using simulated data

The proposed algorithm is tested using a highly harmonic contaminated signal for the three-phase voltage as:

$$v_a(t) = \sin(\omega t - 30^0) + 0.25 \sin(3\omega t) + 0.1 \sin(5\omega t) + 0.05 \sin(7\omega t)$$

The harmonics in other two phases are displaced backward and forward from phase a by 120° and equal in magnitudes, balanced harmonics contamination.

The sampling frequency is chosen to be $F_s = 4. * f_0 * n$, $f_0 = 50$ Hz, where n is the order of harmonic to be identified, n = 1,..,..,N, N is the largest order of harmonics to be expected in the waveform. In this example N=8. A number of sample equals 50 is chosen to estimate the parameters of each harmonic components. Table 3 gives the results obtained when n take the values of 1,3,5,7 for the three phases.

| Harmonic | 1st harmonic | | 3rd harmonic | | 5th harmonic | | 7th harmonic | |
|---|---|---|---|---|---|---|---|---|
| Phase | V | θ | V | θ | V | θ | V | θ |
| A | 1.0 | -30. | 0.2497 | 179.95 | 0.1 | 0.0 | 0.0501 | 0.200 |
| B | 1.0 | -150 | 0.2496 | 179.95 | 0.1 | 119.83 | 0.04876 | -120.01 |
| C | 1.0 | 89.9 | 0.2496 | 179.95 | 0.0997 | -119.95 | 0.0501 | 119.8 |

Table 3. The estimated harmonic in each phase, sampling frequency=1000 Hz and the number of samples=50

Examining this table reveals that the proposed transformation is succeeded in estimating the harmonics content of a balanced three phase system. Furthermore, there is no need to model each harmonic component as was done earlier in the literature. Another test is conducted in this section, where we assume that the harmonics in the three phases are unbalanced. In this test, we assume that the three phase voltages are as follows;

$$v_a(t) = \sin(\omega t - 30^0) + 0.25\sin(3\omega t) + 0.1\sin(5\omega t) + 0.05\sin(7\omega t)$$

$$v_b(t) = 0.9\sin(\omega t - 150^0) + 0.2\sin(3\omega t) + 0.15\sin(5\omega t + 120^0) + 0.03\sin(7\omega t - 120^0)$$

$$v_c(t) = 0.8\sin(\omega t + 90^0) + 0.15\sin(3\omega t) + 0.12\sin(5\omega t - 120^0) + 0.04\sin(7\omega t + 120^0)$$

The sampling frequency used in this case is 1000Hz, using 50 samples. Table 4 gives the results obtained for the positive sequence of each harmonics component including the fundamental component.

| Harmonic | 1st harmonic | | 3rd harmonic | | 5th harmonic | | 7th harmonic | |
|---|---|---|---|---|---|---|---|---|
| Phase | V | θ | V | θ | V | θ | V | θ |
| A | 0.9012 | -29.9 | 0.2495 | 179.91 | 0.124 | 0.110 | 0.0301 | 0.441 |
| B | 0.8986 | -149.97 | 0.2495 | 179.93 | 0.123 | 119.85 | 0.0298 | -120.0 |
| C | 0.900 | 89.91 | 0.2495 | 179.9 | 0.123 | -119.96 | 0.0301 | 119.58 |

Table 4. Estimated positive sequence for each harmonics component

Examining this table reveals that the proposed transformation is produced a good estimate in such unbalanced harmonics for magnitude and phase angle of each harmonics component. In this case the components for the phases are balanced.

## 6.4 Remarks
We present in this section an algorithm to identifying and measuring harmonics components in a power system for quality analysis. The main features of the proposed algorithm are:
- It needs no model for the harmonic components in question.
- It filters out the dc components of the voltage or current signal under consideration.
- The proposed algorithm avoids the draw backs of the previous algorithms, published earlier in the literature, such as FFT, DFT, etc
- It uses samples of the three-phase signals that gives better view to the system status, especially in the fault conditions.
- It has the ability to identify a large number of harmonics, since it does not need a mathematical model for harmonic components.

The only drawback, like other algorithms, if there is a frequency drift, it produces inaccurate estimate for the components under study. Thus a frequency estimation algorithm is needed in this case. Also, we assume that the amplitude and phase angles of each harmonic component are time independent, steady state harmonics identification.

## 7. Fuzzy harmonic components identification

In this section, we present a fuzzy Kalman filter to identify the fuzzy parameters of a general non-sinusoidal voltage or current waveform. The waveform is expressed as a Fourier series of sines and cosines terms that contain a fundamental harmonic and other harmonics to be measured. The rest of the series is considered as additive noise and unmeasured distortion. The noise is filtered out and the unmeasured distortion contributes to the fuzziness of the measured parameters. The problem is formulated as one of linear fuzzy problems. The $n^{th}$ harmonic component to be identified, in the waveform, is expressed as a linear equation: $\underline{A}_{n1}$ $sin(n\omega t) + \underline{A}_{n2} \, cos(n\omega t)$. The $\underline{A}_{n1}$ and $\underline{A}_{n2}$ are fuzzy parameters that are used to determine the fuzzy values of the amplitude and phase of the $n^{th}$ harmonic. Each fuzzy parameter belongs to a symmetrical triangular membership function with a middle and spread values. For example $\underline{A}_{n1} = (p_{n1}, c_{n1})$, where $p_{n1}$ is the center and $c_{n1}$ is the spread. Kalman filtering is used to identify fuzzy parameters $p_{n1}, c_{n1}, p_{n2}$, and $c_{n2}$ for each harmonic required to be identified.

An overview of the necessary linear fuzzy model and harmonic waveform modeling is presented in the next section.

### 7.1 Fuzzy function and fuzzy linear modeling

The fuzzy sets were first introduced by Zadeh [20]. Modeling fuzzy linear systems has been addressed in [8,9]. In this section an overview of fuzzy linear models is presented. A fuzzy linear model is given by:

$$Y = f(x) = \underline{A}_0 + \underline{A}_1 \, x_1 + \underline{A}_2 \, x_2 + \dots + \underline{A}_n \, x_n \tag{66}$$

where $Y$ is the dependent fuzzy variable (output), $\{x_1, x_2, \dots, x_n\}$ set of crisp (not fuzzy) independent variables, and $\{\underline{A}_0, \underline{A}_1, \dots, \underline{A}_n\}$ is a set of symmetric fuzzy numbers. The membership function of $\underline{A}_i$ is symmetrical triangular defined by center and spread values, $p_i$ and $c_i$, respectively and can be expressed as

$$\mu_{A_i}(a_i) = \begin{cases} 1 - \dfrac{|p_i - a_i|}{c_i} & p_i - c_i \le a_i \le p_i + c_i \\[2mm] 0 & \textit{otherwise} \end{cases} \tag{67}$$

Therefore, the function $Y$ can be expressed as:

$$Y = f(x) = (p_0, c_0) + (p_1, c_1) \, x_1 + \dots + (p_n, c_n) \, x_n \tag{68}$$

Where $\underline{A}_i = (p_i, c_i)$ and the membership function of $Y$ is given by:

$$\mu_Y(y) = \begin{cases} 1 - \dfrac{\left|y - \sum\limits_{i=1}^{n} p_i x_i\right|}{\sum\limits_{i=1}^{n} c_i |x_i|} & x_i \ne 0 \\[4mm] 1 & x_i = 0, \quad y = 0 \\[2mm] 0 & x_i = 0, \quad y \ne 0 \end{cases} \tag{69}$$

Fuzzy numbers can be though of as crisp sets with moving boundaries with the following four basic arithmetic operations [9]:

$$[a, b] + [c, d] = [a+c, b+d]$$

$$[a, b] - [c, d] = [a-d, b-c]$$

$$[a, b] * [c, d] = [min(ac, ad, bc, bd), max(ac, ad, bc, bd)]$$

$$[a, b] / [c, d] = [min(a/c, a/d, b/c, b/d), max(a/c, a/d, b/c, b/d)] \tag{70}$$

In the next section, waveform harmonics will be modeled as a linear fuzzy model.

## 7.2 Modeling of harmonics *as a fuzzy model*

A voltage or current waveform in a power system beside the fundamental one can be contaminated with noise and transient harmonics. For simplicity and without loss of generality consider a non-sinusoidal waveform given by

$$v(t) = v_1(t) + v_2(t) \tag{71}$$

where $v_1(t)$ contains harmonics to be identified, and $v_2(t)$ contains other harmonics and transient that will not be identified. Consider $v_1(t)$ as Fourier series:

$$v_1(t) = \sum_{n=1}^{N} V_n \sin(n\omega_0 t + \varphi_n)$$
$$v_1(t) = \sum_{n=1}^{N} [V_n \cos\varphi_n \sin(n\omega_0 t) + V_n \sin\varphi_n \cos(n\omega_0 t)] \tag{72}$$

Where $V_n$ and $\varphi_n$ are the amplitude and phase angle of the $n^{th}$ harmonic, respectively. $N$ is the number of harmonics to be identified in the waveform. Using trigonometric identity $v_1(t)$ can be written as:

$$v_1(t) = \sum_{n=1}^{N} [\overline{A}_{n1} x_{n1} + \underline{A}_{n2} x_{n2}] \tag{73}$$

Where  $x_{n1} = sin(n\omega_0 t)$, $x_{n2} = cos(n\omega_0 t)$        n=1, 2, ..., N
       $\underline{A}_{n1} = V_n \cos\varphi_n$, $\underline{A}_{n2} = V_n \sin\varphi_n$        n=1, 2, ..., N
Now v(t) can be written as:

$$v(t) = \underline{A}_0 + \sum_{n=1}^{N} [\underline{A}_{n1} x_{n1} + \underline{A}_{n2} x_{n2}] \tag{74}$$

Where $\underline{A}_0$ is effective (rms) value of $v_2(t)$.
Eq.(74) is a linear model with coefficients $\underline{A}_0$, $\underline{A}_{n1}$, $\underline{A}_{n2}$, n=1, 2, ..., N. The model can be treated as a fuzzy model with fuzzy parameters each has a symmetric triangular membership function characterized by a central and spread values as described by Eq.(68).

$$v(t) = (p_0 + c_0) + \sum_{n=1}^{N} [(p_{n1} + c_{n1}) x_{n1} + (p_{n1} + c_{n1}) x_{n2}] \tag{75}$$

In the next section, Kalman filtering technique is used to identify the fuzzy parameters. Once the fuzzy parameters are identified then fuzzy values of amplitude and phase angle of each harmonic can be calculated using mathematical operations on fuzzy numbers. If crisp values of the amplitudes and phase angles of the harmonics are required, the defuzzefication is used. The fuzziness in the parameters gives the possible extreme variation that the parameter can take. This variation is due to the distortion in the waveform because of contamination with harmonic components, $v_2(t)$, that have not been identified. If all harmonics are identified, $v_2(t)=0$, then the spread values would be zeros and identified parameters would be crisp rather than fuzzy ones.

Having identified the fuzzy parameters, the $n^{th}$ harmonic amplitude and phase can be calculated as:

$$v_n^2 = \underline{A}_{n1}^2 + \underline{A}_{n2}^2$$
$$\tan \varphi_n = \underline{A}_{n2}^2 \div \underline{A}_{n1}^2 \tag{76}$$

The parameters in Eq. (76) are fuzzy numbers, the mathematical operations defined in Eq. (70) are employed to obtain fuzzy values of the amplitude and phase angle.

### 7.3 Fuzzy amplitude calculation:

Writing amplitude Eq.(76) in fuzzy form:

$$v_n^2 = (p_{v_n^2}, c_{v_n^2}) = (p_{n1}, c_{n1})(p_{n1}, c_{n1}) + (p_{n2}, c_{n2})(p_{n2}, c_{n2}) \tag{77}$$

To perform the above arithmetic operations, the fuzzy numbers are converted to crisp sets of the form $[p_i - c_i, p_i + c_i]$. Since symmetric membership functions are assumed, for simplicity, only one half of the set is considered, $[p_i, p_i + c_i]$. Denoting the upper boundary of the set $p_i + c_i$ by $u_i$, the fuzzy numbers are represented by sets of the form $[p_i, u_i]$ where $u_i > p_i$. Accordingly,

$$v_n^2 = [p_{v_n^2}^2, u_{v_n^2}^2] = [p_{n1}, u_{n1}][p_{n1}, u_{n1}] + [p_{n2}, u_{n2}][p_{n2}, u_{n2}] \tag{78}$$

Then the center and spread values of the amplitude of the $n^{th}$ harmonic are computed as follows:

$$p_{v_n} = \sqrt{p_{v_n^2}} = \sqrt{p_{n1}^2 + p_{n2}^2}$$
$$u_{v_n} = \sqrt{u_{v_n^2}} = \sqrt{u_{n1}^2 + u_{n2}^2} \tag{79}$$
$$c_{v_n} = u_{v_n} - p_{v_n}$$

### 7.4 Fuzzy phase angle calculation

Writing phase angle Eq.(79) in fuzzy form:

$$\tan \varphi_n = (p_{\tan \varphi n}, c_{\tan \varphi n}) = (p_{n2}, c_{n2}) \div (p_{n1}, c_{n1}) \tag{80}$$

Converting fuzzy numbers to sets:

$$\tan \varphi_n = [p_{\tan \varphi n}, u_{\tan \varphi n}] = [p_{n2}, u_{n2}] \div [p_{n1}, c_{n1}] \tag{81}$$

then the central and spread values of the phase angle is given by:

$$p_{\varphi n} = \tan^{-1}(p_{\tan \varphi n}) = \tan^{-1}(p_{n2} / p_{n1})$$
$$u_{\varphi n} = \tan^{-1}(u_{\tan \varphi n}) = \tan^{-1}(u_{n2} / u_{n1}) \tag{82}$$
$$c_{\varphi n} = u_{\varphi n} - p_{\varphi n}$$

## 7.5 Fuzzy modeling for Kalman filter algorithm
### 7.5.1 The basic Kalman filter

The detailed derivation of Kalman filtering can be found in [23, 24]. In this section, only the necessary equation for the development of the basic recursive discrete Kalman filter will be addressed. Given the discrete state equations:

$$x(k+1) = A(k)\, x(k) + w(k)$$

$$z(k) \quad = C(k)\, x(k) + v(k) \tag{83}$$

where

| | | | |
|---|---|---|---|
| $x(k)$ | is | $n \times 1$ | system states. |
| $A(k)$ | is | $n \times n$ | time varying state transition matrix. |
| $z(k)$ | is | $m \times 1$ | vector measurement. |
| $C(k)$ | is | $m \times n$ | time varying output matrix. |
| $w(k)$ | is | $n \times 1$ | system error. |
| $v(k)$ | is | $m \times 1$ | measurement error. |

The noise vectors $w(k)$ and $v(k)$ are uncorrected white noises that have:

Zero means: $\qquad\qquad\qquad\qquad\quad E[w(k)] = E[v(k)] = 0. \tag{84}$

No time correlation: $\quad\;\; E[w(i)\, w^T(j)] = E[v(i)\, v^T(j)] = 0, \;\; for\; i = j. \tag{85}$

Known covariance matrices (noise levels):

$$E[w(k)\; w^T(k)] = Q_1$$

$$E[v(k)\;\; v^T(k)] = Q_2 \tag{86}$$

where $Q_1$ and $Q_2$ are positive semi-definite and positive definite matrices, respectively. The basic discrete-time Kalman filter algorithm given by the following set of recursive equations. Given as priori estimates of the state vector $x^{\wedge}(0) = x^{\wedge}_0$ and its error covariance matrix, $P(0) = P_0$, set $k=0$ then recursively computer:

Kalman gain: $\qquad\quad K(k) = [A(k)\, P(k)\, C^T(k)]\, [C(k)\, P(k)\, C^T(k) + Q_2]^{-1} \tag{87}$

New state estimate:

$$x^{\wedge}(k+1) = A(k)\, x^{\wedge}(k) + K(k)\, [z(k) - C(k)x^{\wedge}(k)] \tag{88}$$

Error Covariance update:

$$P(k+1) = [A(k) - K(k)\, C(k)]\, p(k)\, [A(k) - K(k)\, C(k)]^T + K(k)\, Q_2\, K^T(k) \tag{89}$$

An intelligent choice of the priori estimate of the state $x^{\wedge}_0$ and its covariance error $P_0$ enhances the convergence characteristics of the Kalman filter. Few samples of the output waveform $z(k)$ can be used to get a weighted least squares as an initial values for $x^{\wedge}_0$ and $P_0$:

$$x^{\wedge}_0 = [H^T\, Q_2^{-1}\, H]^{-1}\, H^T\, Q_2^{-1}\; z_0$$

$$P_0 = [H^T Q_2^{-1} H]^{-1} \tag{90}$$

where    $z_0$ is $(m \, m_1) \times 1$ vector of $m_1$ measured samples.
H is $(m \, m_1) \times n$ matrix.

$$z_0 = \begin{bmatrix} z(1) \\ z(2) \\ \vdots \\ z(m_1) \end{bmatrix} \quad and \quad H = \begin{bmatrix} C(1) \\ C(2) \\ \vdots \\ C(m_1) \end{bmatrix} \tag{91}$$

### 7.5.2 Fuzzy harmonic estimation dynamic model
In this sub-section the harmonic waveform is modeled as a time varying discrete dynamic system suited for Kalman filtering. The dynamic system of Eq.(83) is used with the following definitions:
1.    The state transition matrix, $A(k)$, is a constant identity matrix.
2.    The error covariance matrices, $Q_1$ and $Q_2$, are constant matrices.
3.    $Q_1$ and $Q_2$ values are based on some knowledge of the actual characteristics of the process and measurement noises, respectively. $Q_1$ and $Q_2$ are chosen to be identity matrices for this simulation, $Q_1$ would be assigned better value if more knowledge were obtained on the sensor accuracy.
4.    The state vector, $x(k)$, consists of *2N+1* fuzzy parameters.
5.    Two parameters (center and spread) per harmonic to be identified. That mounts to 2N parameters. The last parameter is reserved for the magnitude of the error resulted from the unidentified harmonics and noise. (Refer to Eq. (92)).
6.    $C(k)$ is *3x(2N+1)* time varying measure matrix, which relates the measured signal to the state vector. (Refer to Eq. (106)).
7.    The observation vector, *z(k),* is 3x(2N+1) time varying vector, depends on the signal measurement. (Refer to Eq. (92)).
The observation equation $z(k)=C(k) \, x(k)$ has the following form:

$$\begin{bmatrix} v(k) \\ \delta(k) \\ \eta(k) \end{bmatrix} = \begin{bmatrix} x_{11} & x_{12} & \cdots & x_{N1} & x_{N2} & 0 & 0 & \cdots & 0 & 0 & 0 \\ 0 & 0 & \cdots & 0 & 0 & x_{11} & x_{12} & \cdots & x_{N1} & x_{N2} & 0 \\ 0 & 0 & \cdots & 0 & 0 & 0 & 0 & \cdots & 0 & 0 & 1 \end{bmatrix} \begin{bmatrix} p_{11} \\ p_{12} \\ \vdots \\ p_{N1} \\ p_{N2} \\ c_{11} \\ c_{12} \\ \vdots \\ c_{N1} \\ c_{N2} \\ p_0 \end{bmatrix} \tag{92}$$

Where $x_{n1}$, $x_{n2}$, $n=1, 2, ..., N$ are defined in Eq.(73) with sampling at time instant $t \rightarrow Tk$, $T$ is the signal period and $k = 1, 2, ...$ .

The first row of $C(k)$ is used to identify the center of the fuzzy parameters, while the second row is used to identify the spread parameters. The third raw is used to identify the magnitude of the error produced by the unidentified harmonics and noise. The observation vector $z(k)$ consists of three values. $v(k)$ is the value of the measured waveform signal at sampling instant $k$., $\delta(k)$ and $\eta(k)$ depends on $v(k)$ and the state vector at time instant $k-1$. They are defined below.

Start with $\eta(k)$, it is defined as the square of the error:

$$\eta(k) = e^2 = [v(k) - \hat{v}(k)]^2 \tag{93}$$

$$\hat{v}(k) = \sum_{n=1}^{N} [p_{n1}(k)x_{n1}(k) + p_{n2}(k)x_{n2}(k)] \tag{94}$$

The estimated error in Eq. (93) is computed using the estimated central values of the harmonics of $v_1(t)$ (Ref. Eq. (85)). The reason for estimating the square of the error rather than the error its self is due to the intrinsic nature of the Kalman filter of filtering out any zero means noise.

The second entry of $z(k)$ is $\delta(k)$, which is the measured spread of the identified harmonics,$v_1(t)$ . It can be thought of as $v_2(t)$ modeled as $v_1(t)$ harmonics.

$$\delta(k) = \sum_{n=1}^{N} [c_{n1}(k)x_{n1}(k) + c_{n2}(k)x_{n2}(k] \tag{95}$$

The $x_{n1}$ and $x_{n2}$ are the $v_1(t)$ harmonics and they are well defined at time instant $k$, but $c_{n1}$ and $c_{n2}$ are the measurement error components in the direction of the $n^{th}$ harmonic of $v_1(t)$. They are computed as follows:

$$c_{n1}(k) = e_{peak}(k)\cos(\varphi_n)(k)$$
$$c_{n2}(k) = e_{peak}(k)\sin(\varphi_n)(k) \tag{96}$$

Where $e_{peak}$ is the peak error defined in Eq.(93), $\sin\varphi_n$ and $\sin\varphi_n$ are defined in Eq(87). Since the peak error depends on the measured samples, its mean square is estimated as a separate parameter. It is $p_0(k)$, the last parameter in the state vector. Similarly, $\cos\varphi_n$ and $\sin\varphi_n$ are unknown parameters that are estimated in the state vector. $e_{peak}$, $\cos\varphi_n$ and $\sin\varphi_n$ are computed as follows:

$$e_{peak}(k) = [2p_0(k-1)]^{1/2}$$
$$\cos\varphi_n(k) = p_{n1}(k-1) / [p_{n1}^2(k-1) + p_{n2}^2(k-1)]$$
$$\cos\varphi_n(k) = p_{n2}(k-1) / [p_{n1}^2(k-1) + p_{n2}^2(k-1)] \tag{97}$$

## 7.5.3 Simulation results

To verify the effectiveness of the proposed harmonic fuzzy parameter identification approach, simulation examples are given below.

### 7.5.4 One harmonic identification

As a first example consider identification of one harmonic only, N=1. Consider a voltage waveform that consists of two harmonics, one fundamental at 50Hz and a sub-harmonic at 150Hz which is considered as undesired distortion contaminating the first harmonic.

$$v(t) = 1.414\sin(100\pi t + \pi / 6)$$
$$+0.3\sin(300\pi t + \pi / 5) \tag{98}$$

For parameter estimation using Kalman filter, the voltage signal is sampled at frequency 1250Hz and used as measurement samples. Converting Eq (88) to discrete time, t → 0.08k, where k is the sampling time, and using the notation of Eq. (71), $v_1(k)$ and $v_2(k)$ are defined as:

$$v_1(k) = 1.414\sin(0.08\pi k + \pi / 6)$$
$$v_2(k) = 0.30\sin(0.24\pi k + \pi / 5) \tag{99}$$

using the notation of Eq.(88), the time fuzzy model is given by:

$$v(k) = \underline{A}_o + \underline{A}_{11}\, x_{11}(k) + \underline{A}_{12}\, x_{12}(k)$$

where $x_{11}(k) = \sin(0.08\pi k)$, $x_{12}(k) = \cos(0.08\pi k)$ and the parameters to be identified are: $\underline{A}_o = (p_o, 0)$, $\underline{A}_{11} = (p_{11}, c_{11})$ and $\underline{A}_{12} = (p_{12}, c_{12})$. The observation equation, $z(k) = C(k)\, x(k)$, becomes:

$$z(k) = \begin{bmatrix} v \\ \delta \\ \eta \end{bmatrix} = \begin{bmatrix} x_{11} & x_{12} & 0 & 0 & 0 \\ 0 & 0 & x_{11} & x_{12} & 0 \\ 0 & 0 & 0 & 0 & 1 \end{bmatrix} \begin{bmatrix} p_{11} \\ p_{12} \\ c_{11} \\ c_{12} \\ p_0 \end{bmatrix} \tag{100}$$

The argument (k) of all variables in Eq.(100) has been omitted for simplicity of notation. With initial state vector $x(0) = [1\ 1\ 1\ 1\ 1]^T$ the following estimated parameters are obtained:
$\underline{A}_0 = (0.052, 0.0)$
$\underline{A}_{11} = (1.223, 0.330)$
$\underline{A}_{12} = (0.710, 0.219)$
Computing the amplitude and phase:
$V_1 = (1.414, 0.395)$
$\varphi_1 = (0.166\pi, 0.014)$
Figures 52 and 53 show the convergence of the center and spread of the first harmonic parameters, respectively.
Figure (54) shows the measured $v(t)$ and estimated (crisp) first harmonic, while Figure (55) illustrates the estimated fuzziness of $v(t)$ by reconstructing waveforms of the form.

$$v_{pc}(t) = (p_{11} \pm c_{11})\, x_{11} + (p_{12} \pm c_{12})\, x_{12} \tag{101}$$

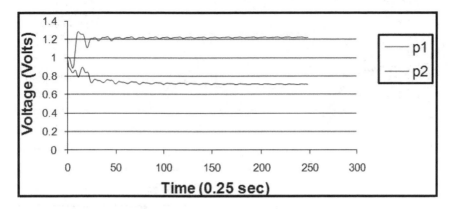

Fig. 52. First Harmonic Centre Paramaters.

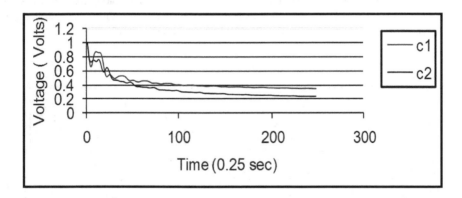

Fig. 53. First Harmonic spread parameters.

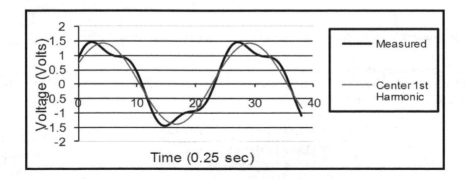

Fig. 54. Mauserd waveform and estimated central of the first harmonic.

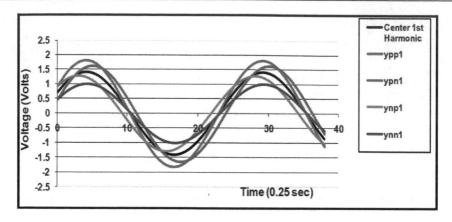

Fig. 55. 1 st Harmonic with its fuzzy variations.

Figure (87) shows *v(t)* together with maximum and minimum possible variation (fuzzy) v(t) can take. It can be observed that the measured v(t) is within the estimated fuzziness and that the extreme fuzzy variations is shaped up according to the measured v(t).

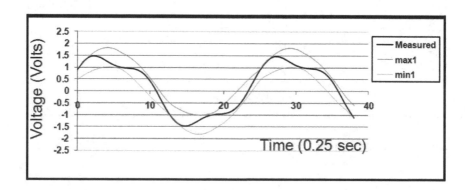

Fig. 56. Mauserd waveform and ist maximum and minimum fuzzy variation

### 7.5.5 Two harmonics identification
Next, consider identifying four harmonics at 50Hz, 100Hz, 150 Hz and 200Hz.The voltage waveform is given in Eq.(102).

$$
\begin{aligned}
v(t) &= 1.414\sin(100\pi t + 0.16667\pi) \\
&+ 1.0\sin(200\pi t + 0.26667\pi) \\
&+ 0.3\sin(300\pi t + 0.2\pi) \\
&+ 0.1\sin(400\pi t + 0.35\pi)
\end{aligned}
\tag{102}
$$

Then, for estimating the first two harmonics and using Eq.(71) $v_1(k)$ and $v_2(k)$ are obtained as follows:

$$v_1(k) = 1.414\sin(0.08\pi k + 0.16667\pi)$$
$$+1.0\sin(0.16\pi k + 0.26667\pi) \tag{103}$$
$$v_2(k) = 0.3\sin(0.24\pi k + 0.2\pi)$$
$$+0.1\sin(0.32\pi k + 0.35\pi)$$

And the linear fuzzy model is given by:

$$v(k) = A_o + A_{11} x_{11}(k) + A_{12} x_{12}(k) + A_{21} x_{21}(k) + A_{22} x_{22}(k)$$

Where $x_{11}(k)=\sin(0.08\pi k)$, $x_{12}(k)=\cos(0.08\pi k)$, $x_{21}(k)=\sin(0.16\pi k)$, $x_{22}(k)=\cos(0.16\pi k)$.
Therefore, there are nine parameters to be estimated and their estimated values are found to be:

$A_o =$     (0.058,    0.0)
$A_{11}=$     (1.224,    0.330)
$A_{12}=$     (0.707,    0.219)
$A_{21}=$     (0.669,    0.267)
$A_{22}=$     (0.743,    0.307)
Computing the amplitude and phase:
$V_1=$     (1.414, 0.395)
$\varphi_1=$     (0.166π, 0.014)
$V_2=$     (1.00, 0.406)
$\varphi_2=$     (0.266π, 0.005)
Figures (57-59) show the crisp and fuzzy variations of v(t).

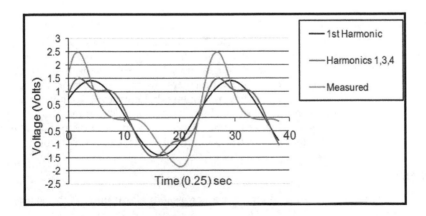

Fig. 57. Efect of removing 2nd Harmonic

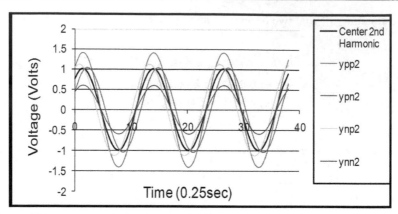

Fig. 58. Second Harmonic with its fuzzy variation

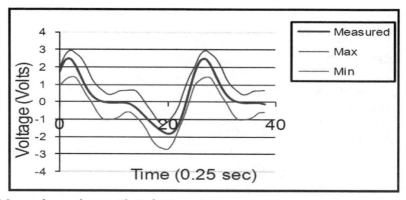

Fig. 59. Measurd waveform with its fuzzy variations

### 7.5.6 Conclusion and remarks

In this paper, the harmonics of a non-sinusoidal waveform is identified. The approach is based on fuzzy Kalman filtering. The basic idea is to identify fuzzy parameters rather than crisp parameters. The waveform is written as a linear model with fuzzy parameters from which the amplitude and phase of the harmonics are measured. Kalman filter is used to identify the fuzzy parameters. Each fuzzy parameter belongs to a triangular symmetric membership function consisting of center and spread values. Obtaining fuzzy parameters rather than crisp ones yields all possible extreme variations the parameters can take. This is useful in designing filters to filter out undesired harmonics that cause distortion.

## 8. References

J. Arrillaga, D.A. Bradley and P.S. Bodger, "Power System Harmonics," John Wiley & Sons, New York, 1985.

IEEE Working Group on Power System Harmonics, "Power System Harmonics: An Overview," IEEE Trans. on Power Apparatus and Systems, Vol. PAS-102, No. 8, pp. 2455-2460, August 1983.

S.A. Soliman, G.S. Christensen, D.H. Kelly and K.M. El-Naggar, "A State Estimation Algorithm for Identification and Measurement of Power System Harmonics," Electrical Power System Research Jr., Vol. 19, pp. 195-206, 1990.

M.S. Saddev and M. Nagpal, "A Recursive Least Error Squares Algorithm for Power System Relaying and Measurement Applications," IEEE Trans. on Power Delivery, Vol. 6, No. 3, pp. 1008-1015, 1991.

S.A. Soliman, K. El-Naggar and A. Al-Kandari, "Kalman Filtering Algorithm for Low Frequency Power Systems Sub-harmonics Identification," Int. Jr. of Power and Energy Systems, Vol. 17, No. 1, pp. 38-43, 1998.

E.A. Abu Al-Feilat, I. El-Amin and M. Bettayeb, "Power System Harmonic Estimation: A Comparative Study," Electric Power Systems Research, Vol. 29, pp. 91-97, 1991.

A.A. Girgis, W.B. Chang and E.B. Markram, "A Digital Recursive Measurement Scheme for On-Line Tracking of Power System Harmonics," IEEE Trans. on Power Delivery, Vol. 6, No. 3, pp. 1153-1160, 1991.

H.M. Beides and G.T. Heydt, "Dynamic State Estimation of Power System Harmonics Using Kalman Filter Methodology," IEEE Trans. on Power Delivery, Vol. 6, No. 4, pp. 1663-1670, 1991.

H. Ma and A.A. Girgis, "Identification and Tracking of Harmonic Sources in a Power System Using a Kalman Filter," IEEE Trans. on Power Delivery, Vol. 11, No. 3, pp. 1659-1665, 1998.

V.M.M. Saiz and J. Barros Gaudalupe, "Application of Kalman Filtering for Continuous Real-Time Traching of Power System Harmonics," IEE Proc.-Gener. Transm. Distrib. Vol. 14, No. 1, pp. 13-20, 1998.

S.A. Soliman and M.E. El-Hawary, "New Dynamic Filter Based on Least Absolute Value Algorithm for On-Line Tracking of Power System Harmonics," IEE Proc.-Generation, Trans. Distribution., Vol. 142, No. 1, pp. 37-44, 1005.

S.A. Soliman, K. El-Naggar, and A. Al-Kandari, "Kalman Filtering Based Algorithm for Low Frequency Power Systems Sub-harmonics Identification," Int. Jr. of Power and Energy Systems, Vol. 17, No. 1, pp. 38-43, 1998.

A. Al-Kandari, S.A. Soliman and K. El-Naggar, "Digital Dynamic Identification of Power System Sub-harmonics Based on Least Absolute Value," Electric Power Systems Research, Vol. 28, pp. 99-104, 1993.

A. A. Girgis and J. Qiu, Measurement of the parameters of slowly time varying high frequency transients, IEEE Trans. On Inst. And Meas., 38(6) (1989) 1057-1062.

A. A. Girgis, M. C. Clapp, E. B. Makram, J. Qiu, J. G. Dalton and R. C. Satoe, Measurement and characterization of harmonic and high frequency distortion for a large industrial load, IEEE Trans. Power Delivery, 5(1) (1990) 427-434

A. A. Girgis, W. Chang, and E. B. Makram, Analysis of high-impedance fault generated signals using a Kalaman filtering approach, IEE Trans. On Power Delivery, 5(4) (1990).

S. A. Soliman, K. El-Naggar, and A. Al-Kandari, Kalman filtering based algorithm for low frequency power systems sub-harmonics identification, International Journal of Power and Energy Systems 17(1), (1997) 38-42.

S. A. Soliman and M. E. El-Hawary, Application of Kalaman filtering for online estimation of symmetrical components for power system protection, Electric Power Systems Research 38 (1997) 113-123.

S.A. Soliman, I. Helal, and A. M. Al-Kandari, Fuzzy linear regression for measurement of harmonic components in a power system, Electric Power System Research 50 (1999) 99-105.

L.A. Zadeh, Fuzzy sets as a basis for theory of possibility, Fussy Sets and Systems, Vol. 1, pp 3-28, 1978.

H. Tanaka, S. Vejima, K. Asai, Linear regression analysis with fussy model, IEEE Trans. On System, Man, and Cybernetics, Vol. 12, No. 6, pp 903-907, 1982.

Timothy J. Ross, Fuzzy logic with engineering applications, McGraw Hill, 1995.

R. G. Brown, Introduction to random signal analysis and Kalman filtering, New York: John Wiley and Sons, 1983.

G. F. Franklin, J. D. Powel and M. L. Workman, Digital control of dynamic system, 2nd edition, Addison Wesley, 1990.

S. K. Tso and W. L. Chan, "Frequency and Harmonic Evaluation Using Non-Linear Least Squares Techniques" Jr. of Electrical and Electronic Engineers., Australia , Vol. 14, No. 2, pp. 124-132, 1994.

M. M. Begovic, P. M. Djuric S. Dunlap and A. G. Phadke, "Frequency Tracking in Power network in the Presence of Harmonics" IEEE Trans. on Power Delivery, Vol. 8, No. 2, pp. 480-486, 1993.

S. A. Soliman, G. S. Christensen, and K. M. El-Naggar,  "A New Approximate Least Absolute Value Based on Dynamic Filtering for on-line Power System Frequency Relaying", Elect. Machines & Power Systems, Vol. 20, pp. 569-592, 1992.

S. A. Soliman, G. S. Christensen, D. H. Kelly, and K. M. El-Naggar, "Dynamic Tracking of the Steady State Power System Magnitude and Frequency Using Linear Kalman Filter: a Variable Frequency Model", Elect. Machines & Power Systems, Vol. 20, pp. 593-611, 1992.

S. A. Soliman and G. S. Christensen, "Estimating of Steady State Voltage and Frequency of Power Systems from Digitized Bus Voltage Samples", Elect. Machines & Power Systems, Vol. 19, pp. 555-576, 1991.

P. K. Dash, D. P. Swain, A. C. Liew, and S. Rahman, "An Adaptive Linear Combiner for on-line Tracking of Power System Harmonics", IEEE Trans. on Power Systems, Vol.114, pp. 1730-1735, 1996.

A. Cavallini and G. C. Montanari, "A Deterministic/Stochastic Framework for Power System Harmonics Modeling", IEEE Transaction on Power Systems, Vol.114, 1996.

S. Osowski, "Neural Network for Estimation of Harmonic Components in a Power System", IEE Proceeding-C, Vol.139, No.2, pp.129-135, 1992.

S. Osowski, "SVD Technique for Estimation of Harmonic Components in a Power System: a Statistical Approach", IEE Proceedings, Gen. Trans. & Distrb., Vol. 141, No.5, pp.473-479, 1994.

S. A. Soliman , G. S. Christensen, D.H. Kelly, and K. M. El-Naggar, " Least Absolute Value Based on Linear Programming Algorithm for Measurement of Power System Frequency from a Distorted Bus Voltage Signal", Elect. Machines & Power Systems, Vol. 20, No. 6, pp. 549-568, 1992.

P. J. Moore, R. D. Carranza and A. T. Johns, " Model System Tests on a New Numeric Method of Power System Frequency Measurement" IEEE Transactions on Power Delivery, Vol. 11, No. 2, pp.696-701, 1996.

P. J. Moore, J. H. Allmeling and A. T. Johns, " Frequency Relaying Based on Instantaneous Frequency Measurement", IEEE   Transaction on Power Delivery, Vol. 11, No. 4, pp.1737-1742, 1996

T. Lobos and J. Rezmer, "Real -Time Determination of Power System Frequency", IEEE Transaction on Instrumentation and Measurement, Vol. 46, No. 4, pp.877-881, 1998.

T. S. Sidhu, and M.S. Sachdev, "An Iterative Techniques for Fast and Accurate Measurement of Power System Frequency", IEEE Transaction on Power Delivery, Vol. 13, No. 1, pp.109-115, 1998.

J. Szafran, and W.,"Power System Frequency Estimation", IEE Proc.-Genre. Trans., Distrib., Vol. 145, No. 5, pp.578-582, 1998.

T. S. Sidhu, " Accurate Measurement of Power System Frequency Using a Digital Signal Processing Technique", IEEE Transaction on Instrumentation and Measurement, Vol. 48, No. 1 , pp.75-81, 1999.

P. K. Dash, A. K. Pradhan, and G. Panda, " Frequency Estimation of Distorted Power System Signals Using Extended Complex Kalman Filter", IEEE Transaction on Power Delivery, Vol. 14, No. 3, pp.761- 766,1999.

S.A. Soliman, H. K Temraz and M. E. El-Hawary, "Estimation of Power System Voltage and Frequency Using the Three-Phase Voltage Measurements and $\alpha\beta$–Transformation", Proceeding of Middle East Power System Conference, MEPCON`2000, Cairo, Ain Shams University, March 2000.

M. E. El-Hawary, "Electric Power Applications of Fuzzy Systems", IEEE Press, Piscataway, NJ, 1998.

Quanming Zhang, Huijin Liu, Hongkun Chen, Qionglin Li, and Zhenhuan Zhang," A Precise and Adaptive Algorithm for Interharmonics Measurement Based on Iterative DFT", IEEE TRANSACTIONS ON POWER DELIVERY, VOL. 23, NO. 4, OCTOBER 2008.

Walid A. Omran, Hamdy S. K. El-Goharey, Mehrdad Kazerani, and M. M. A. Salama," Identification and Measurement of Harmonic Pollution for Radial and Nonradial Systems", IEEE TRANSACTIONS ON POWER DELIVERY, VOL. 24, NO. 3, JULY 2009

Ekrem Gursoy, and Dagmar Niebur," Harmonic Load Identification Using Complex Independent Component Analysis ",IEEE TRANSACTIONS ON POWER DELIVERY, VOL. 24, NO. 1, JANUARY 2009

Jing Yong, Liang Chen, and Shuangyan Chen," Modeling of Home Appliances for Power Distribution System Harmonic Analysis", IEEE TRANSACTIONS ON POWER DELIVERY, VOL. 25, NO. 4, OCTOBER 2010

Elcio F. de Arruda, Nelson Kagan, and Paulo F. Ribeiro," Harmonic Distortion State Estimation Using an Evolutionary Strategy", IEEE TRANSACTIONS ON POWER DELIVERY, VOL. 25, NO. 2, APRIL 2010

Mohsen Mojiri, Masoud Karimi-Ghartemani and Alireza Bakhshai," Processing of Harmonics and Interharmonics Using an Adaptive Notch Filter", IEEE TRANSACTIONS ON POWER DELIVERY, VOL. 25, NO. 2, APRIL 2010

Cong-Hui Huang, Chia-Hung Lin, and Chao-Lin Kuo,' Chaos Synchronization-Based Detector for Power-Quality Disturbances Classification in a Power System" IEEE TRANSACTIONS ON POWER DELIVERY, VOL. 26, NO. 2, APRIL 2011

Gary W. Chang, Shin-Kuan Chen, Huai-Jhe Su, and Ping-Kuei Wang," Accurate Assessment of Harmonic and Interharmonic Currents Generated by VSI-Fed Drives Under Unbalanced Supply Voltages" IEEE TRANSACTIONS ON POWER DELIVERY, VOL. 26, NO. 2, APRIL 2011

Abner Ramirez," The Modified Harmonic Domain: Inter-harmonics" IEEE TRANSACTIONS ON POWER DELIVERY, VOL. 26, NO. 1, JANUARY 2011

Hooman E. Mazin, Wilsun Xu, and Biao Huang," Determining the Harmonic Impacts of Multiple Harmonic-Producing Loads" IEEE TRANSACTIONS ON POWER DELIVERY, VOL. 26, NO. 2, APRIL 2011.

# 2

# Voltage Harmonics Measuring Issues in Medium Voltage Systems

Jarosław Łuszcz
*Gdańsk University of Technology*
*Poland*

## 1. Introduction

Voltage harmonic distortion level is one of the significant parameters of power quality in power system. Numerous problems related to voltage and current harmonic effects for contemporary power systems are commonly observed nowadays. Levels and spectral content of voltage distortions injected into electric power grids are tending to increase despite the fact that the acceptable levels are determined by numerous regulations. Voltage distortion assessments, especially in middle and high voltage grids, are usually based on measurements in which voltage transformers are commonly used. The transfer ratio of a voltage transformer fed by distorted primary voltage with harmonic components of frequency higher than fundamental can be different for high frequency components in comparison with the fundamental frequency.

During the last decades primary problems related to voltage distortions have been usually encountered in frequency range up to 40th harmonic, mostly in LV grids. Nowadays, due to the evident increase of the overall power of nonlinear power electronic loads connected to grid and higher modulation frequencies widely used, distorted voltage propagates deeply into MV grids and goes evidently beyond frequency of 2 kHz.

This chapter presents problems of voltage harmonic transfer accuracy through voltage transformers which are usually used for power quality monitoring in medium and high voltage grids (Kadar at al., 1997, Seljeseth at al., 1998, Shibuya at al., 2002, Mahesh at al., 2004, Yao Xiao at al., 2004, Klatt at al., 2010). A simplified lumped parameters circuit model of the voltage transformer is proposed and verified by simulation and experimental investigations. A number voltage transformers typically used in medium voltage grid have been tested in the conducted disturbances frequency range up to $30\,MHz$. The obtained results prove that broadband voltage transfer function of the voltage transformer usually exhibits various irregularities, especially in high frequency range, which are primarily associated with windings' parasitic capacitances.

Frequency dependant voltage transfer characteristic of voltage transformer induces extra measurement errors which have to be taken into account in order to achieve desired final relatively high accuracy required for power quality monitoring systems.

## 2. Circuit modelling of voltage transformers

Classical voltage transformer (VT) is a two or three winding transformer with a relatively high transformation ratio and low rated power, intended to supply only measuring inputs of metering apparatus or protection relays extensively used in power system. VT are mostly used in medium voltage (MV) and high voltage (HV) systems for separation of the measuring and protecting circuit from high voltage hazard.

Rated primary voltages of VTs, typically used in power system, have to correspond to rated voltages of MV and HV transmission lines in particular power system. Secondary rated voltage levels usually used in a typical measuring and protection systems are: *100 V, 100/3 V, 100/√3 V* what results with transformation ratios of the order from few tenth up to few hundredths for MV VT and more than thousand for HV VT. Such a high transformation ratio and low rated power of VT influence significantly its specific parameters, especially related to performance in wide frequency range.

The classical equivalent circuit model of two windings transformer widely used for modelling VT for power frequency range is presented in Fig.1. This model consists of leakage inductances of primary winding $L_p$ and secondary winding $L_s$ and magnetizing inductance $L_m$. Corresponding resistances represent VT losses in magnetic core $R_m$ and windings $R_p$, $R_s$.

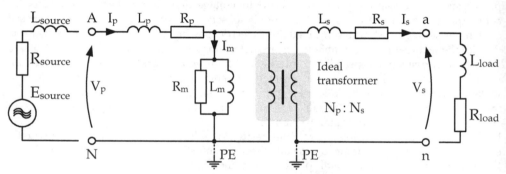

Fig. 1. Classical equivalent circuit model of a voltage transformer

For VT operated under power frequency and rated load presented circuit model can be simplified radically because magnetizing inductance $L_m$. is usually many times higher than leakage inductances $L_m >> L_p$, $L_s$ and VT nominal load impedance $Z_{load} = R_{ld} + j\omega L_{load}$ is usually much higher than secondary leakage impedance $Z_s = R_s + j\omega L_s$ ($Z_{ld} >> Z_s$). This assumption cannot be adopted for frequencies varying far from power frequency range because VT reactance change noticeably with frequency what results with VT transformation ratio change.

Based on this model, which characterizes two not ideally coupled inductances, frequency dependant transfer characteristic for frequencies higher than the nominal (50 or 60 Hz) can be estimated as well. Theoretical wideband transfer characteristic of VT modelled by using classic circuit model is presented in Fig.2 where low corner frequency of pass band $f_{low}$ and high corner frequency of pass band $f_{high}$ can be defined based on *3 dB* transfer ratio decrease margin assumption.

Low and high frequency response of VT can be determined analytically based on VT classic circuit model parameters. For wideband analysis simplification classical circuit model of VT can be represented as a serial connection of high pass filter (HPF), ideal transformer and low pass filter (LPF) (Fig.3). According to this simplification, the pass band characteristic of high pass $LC$ filter is mainly correlated to VT primary side parameters ($R_{HPF}$, $L_{HPF}$) and the pass band characteristic of low pass $LC$ filter is mainly correlated to parameters of secondary side ($R_{LPF}$, $L_{LPF}$).

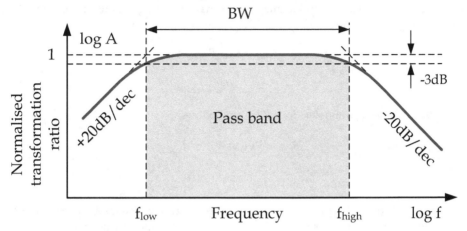

Fig. 2. Theoretical transfer ratio wideband characteristic of VT modelled by classical circuit model

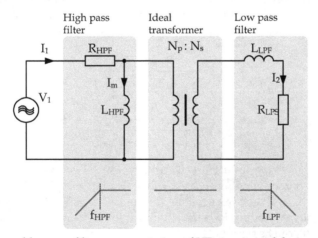

Fig. 3. High pass and low pas filter representation of VT circuit model

Based on this assumption the low corner frequency $f_{low}$ of VT transfer characteristic can be easily defined by formula (1) and high corner frequency $f_{high}$ by formula (2).

$$f_{low} = f_{HPF} = \frac{R_{HPF}}{2\pi L_{HPF}} \tag{1}$$

$$f_{high} = f_{LPF} = \frac{R_{LPF}}{2\pi L_{LPF}} \tag{2}$$

Wideband analysis of VT transfer characteristic requires taking into account also external impedances of measured voltage source and VT load. Therefore, equivalent resistance of high pass filter $R_{HPF}$ can be defined as a sum of VT primary winding resistance and primary voltage source resistance $R_{source}$ (3). Respectively high pass filter equivalent inductance $L_{HPF}$ is a sum of VT magnetizing inductance $L_m$, primary winding leakage inductance $L_p$ and primary voltage source inductance $L_{source}$ (4).

$$R_{HPF} = R_p + R_{source} \tag{3}$$

$$L_{HPF} = L_p + L_m + L_{source} \tag{4}$$

Analogous equivalent parameters for LPF are as follows:

$$R_{LPF} = R_p + \eta^2 R_s + \eta^2 R_{load} \tag{5}$$

$$L_{LPF} = L_p + \eta^2 L_s + \eta^2 L_{load} \tag{6}$$

For RL type low and high pass filters 3 dB pass band margin is obtained for the frequency at which magnitudes of filter resistance $R$ is equal to magnitude of filter reactance $X_L=2\pi fL$. According to this formula, the corner frequency of low pass equivalent filter $f_{LPF}$ determines the lowest signal frequency $f_{low}$ transformed by VT (7) and the corner frequency of high pass equivalent filter $f_{HPF}$ determines the highest signal frequency $f_{high}$ transformed by VT (8), where $\eta=N_P/N_S$ is a VT winding ratio.

$$f_{low} = \frac{R_p + R_{source}}{2\pi \left(L_p + L_m + L_{source}\right)} \tag{7}$$

$$f_{high} = \frac{R_p + \eta^2 R_s + \eta^2 R_{load}}{2\pi \left(L_p + \eta^2 L_s + \eta^2 L_{load}\right)} \tag{8}$$

Assuming that magnetizing inductance $L_m$ of a typical VT is much higher than leakage inductance $L_m$ of primary winding ($L_m \gg L_p$) and also much higher than primary voltage source inductance ($L_m \gg L_{source}$) the equation (7) can be simplified to (9). Similarly, because resistance of secondary winding $R_s$ is usually much lower than load resistance $R_{load}$ ($R_s \ll R_{load}$) the equation (8) can be simplified to (10).

$$f_{low} \approx \frac{R_p + R_{source}}{2\pi \cdot L_m} \tag{9}$$

$$f_{high} \approx \frac{\eta^2 R_{load}}{2\pi \left(L_p + \eta^2 L_s + \eta^2 L_{load}\right)} \tag{10}$$

Finally, VT bandwidth $BW$ can be estimated by using formula (11) and relative bandwidth $BW \, [\%]$ using formula (12).

$$BW = f_{high} - f_{low} = \frac{\eta^2 R_{Load}}{L_p + \eta^2 L_s + \eta^2 L_{Load}} - \frac{R_p + R_{source}}{L_m} \tag{11}$$

$$BW[\%] = 2 \cdot \frac{f_{high} - f_{low}}{f_{high} + f_{low}} = 2 \cdot \frac{\eta^2 L_m R_{load} - \left(L_p + \eta^2 L_s + \eta^2 L_{Load}\right)\left(R_p + R_{source}\right)}{\eta^2 L_m R_{load} + \left(L_p + \eta^2 L_s + \eta^2 L_{Load}\right)\left(R_p + R_{source}\right)} \tag{12}$$

Summarizing, based on classical circuit model analysis, low frequency response of VT is mostly dependant on its leakage to magnetizing impedance ratio which limits transfer characteristic in low frequency range. In applications where voltage source impedance is relatively high and cannot be neglected an extra pass band limitation is observed due to its influence (Fig.4). High frequency response of VT is dependant mainly on its leakage impedance to load impedance ratio which limits transfer characteristic in high frequency range. Increase of load impedance extends bandwidth of VT towards higher frequencies. Theoretically, for very low VT load its pass band can be very wide from the magnetic coupling point of view, nevertheless parasitic capacitances usually limit noticeably VT pass band in high frequency range.

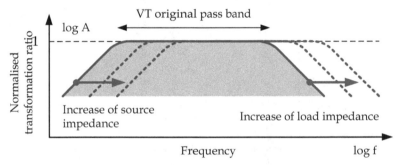

Fig. 4. Influence of source and load impedance on the VT pass band

## 3. Broadband modelling of voltage transformers

Modelling of VT in a wide frequency range using classical circuit model is usually not adequate enough. The foremost reasons for the inadequacy of the classical circuit model are the parasitic capacitances of windings and frequency dependant voltage source and VT load impedances. Parasitic capacitances existing in windings change noticeably the transformation ratio characteristic, particularly in high frequency range.

Parasitic capacitance is an effect of proximity of windings and its sections to each other and to other conductive usually grounded elements, like for example magnetic core, electric shields and other conductive elements of VT. Parasitic capacitances of VT windings are usually unwanted and unluckily unavoidable; there are only various techniques used to reduce its values or change distribution. Parasitic capacitances of VT change radically its behaviour in high frequency range usually reduce evidently the pass band bandwidth with flat transfer characteristic. Parasitic capacitances of VT windings have distributed nature strictly correlated with particular winding arrangement, therefore their identification and modelling is problematic (Vermeulen at al., 1995, Islam at al., 1997, Luszcz, 2004a, 2004b,

Mohamed at al., 2008). Consequences of parasitic capacitances are especially significant for multilayer windings with high number of turns which is characteristic for high voltage and low power transformers like VT. The most essential categories of partial parasitic capacitances occurring in typical VT windings are presented in Fig. 5.

Identification of not equally distributed partial parasitic capacitances for particular VT require detailed specification of winding arrangement, is extremely elaborative and usually do not provide adequate enough results. Difficulties of parasitic capacitances identification can be reduced by defining lumped equivalent capacitances which represent groups of many partial capacitances related to entire winding or part of windings; for example single layer of winding. Noticeable simplification of parasitic capacitances distribution in VT winding can also be achieved by changing winding arrangement and introducing windings' shields. Example of influence of windings' shields on parasitic capacitances distribution is presented in Fig. 6 where the inter-winding capacitance, usually most noticeable, is radically reduced by introducing additional winding-to-shield capacitances.

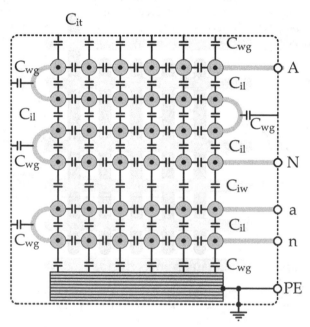

Fig. 5. Major categories of parasitic capacitances occurring in VT windings: $C_{wg}$ – winding to ground, $C_{il}$ – interlayer, $C_{it}$ – inter-turn, $C_{iw}$ – inter-winding

Use of lumped representation of parasitic capacitances allows reducing winding model complexity and consequently simplifies noticeably its parameters identification process. A possible to apply winding model simplification level should be closely correlated with the expected adequacy in a given frequency range and depends evidently on particular winding arrangement complexity. Commonly, three methods of winding parasitic capacitances circuit representations are used to model transformer windings (Fig. 7):

- winding terminals related – where all defined lumped equivalent capacitances are connected to windings' terminals only,

- partially distributed – lumped parasitic capacitances are specified for most representative internal parts of winding, like for example windings layers, winding shields,
- fully distributed – windings are modelled as a series and parallel combination of inductances and capacitances which form ladder circuit with irregular parameter distribution.

Generally more detailed parasitic capacitances representation allows obtaining higher level of model adequacy in wider frequency range. Nevertheless because of identification problems the model complexity should be kept within a reasonable level to allow achieving high usefulness.

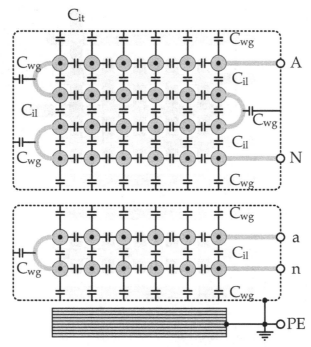

Fig. 6. Parasitic capacitances arrangement in shielded VT windings

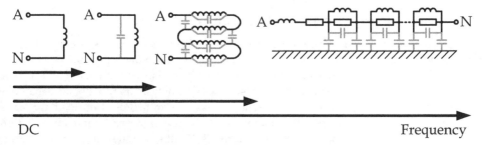

Fig. 7. Typical circuit representations of winding parasitic capacitances

The influence of winding parasitic capacitances on the VT transfer ratio also depends on winding grounding method used in measuring application. In Fig. 8 two mostly used VT winding configurations are presented, where primary winding is connected to measured voltage in a different way. VTs configured as one-side grounded primary and secondary windings are commonly used for phase voltage measurement in power system, while VT floating primary winding allows for direct measurement of inter-phase voltages (Fig.9). The analysis of VT transfer characteristic for VT with both windings grounded is noticeably simpler, therefore presented further analysis based on the proposed circuit model and experimental tests have been limited to this case.

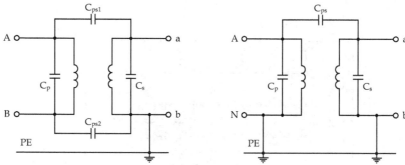

Fig. 8. Different configurations of VT primary winding grounding

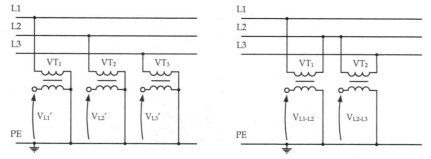

Fig. 9. Typical VT connection for phase and inter-phase voltage measurement in tree phase power system

## 4. Wideband parameters identification of voltage transformer

Particular voltage transformation ratio of VT in high frequency range is closely related to impedance–frequency characteristics of primary and secondary windings. Therefore measurement results of VT magnetizing and leakage impedances within the investigated frequency range are the fundamental data resources for analysis its broadband behaviour. Measurement of VT impedances can be done similarly as in a typical no load and short circuit tests recommended for power frequency with the use of sweep frequency excitation. Examples of magnitude and phase characteristics of the magnetizing impedance of investigated VT are presented in Fig. 10 and leakage impedance in Fig. 11.

Concise analysis of the exemplary magnetizing and leakage impedance characteristics allow estimating characteristic frequency ranges correlated with the VT different performance. Firstly, measured impedances exhibit characteristic serial resonances which appear in frequency range around $f_{r1} \approx 150\,Hz$ and around $f_{r2} \approx 200\,kHz$ for magnetizing and leakage impedances respectively. On the magnetizing impedance characteristic, between these two frequencies $f_{r1}$ and $f_{r2}$, serial resonance is clearly visible for the frequency range around $f_{r3} \approx 2\,kHz$. These three particular resonance frequencies $f_{r1}$, $f_{r2}$, $f_{r3}$ divide the frequency spectrum into three sub-ranges closely related to VT behaviour. Secondly, in the frequency range between around $3\,kHz$ and $20\,kHz$ several less meaningful resonances can be observed which are correlated with local resonances appearing in winding internal subsections.

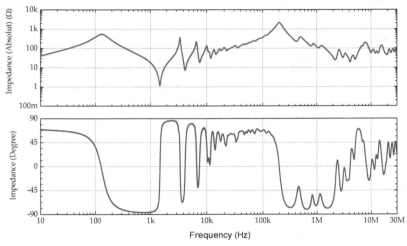

Fig. 10. Measured magnetizing impedance-frequency characteristics of investigated VT

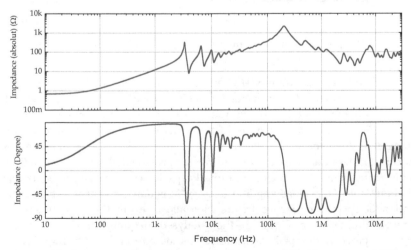

Fig. 11. Measured leakage impedance-frequency characteristics of investigated VT

Simplified representation of VT magnetizing and leakage impedances is presented in Fig. 13, where characteristic resonance frequencies $f_{r1}$, $f_{r2}$, $f_{r3}$ and related to them frequency bands $B_1$, $B_2$, and $B_3$ are emphasized. The frequency band $B_1$ below $f_{r3}$ can be characterized as a VT pass band where the magnetizing impedance is much higher than the leakage impedance; therefore in this frequency range between primary and secondary windings magnetic coupling effect is dominating. In the frequency band $B_2$, between $f_{r2}$ and $f_{r3}$, the magnetizing impedance values are comparable to the leakage impedance, what weakening noticeably the magnetic coupling effect and the influence of VT load impedance on the transfer ratio characteristic became significant. In the frequency range above $f_{r2}$ (band $B_3$) the capacitive character of magnetizing and leakage impedances of VT is dominating, what means that capacitive type of coupling between VT windings is predominant.

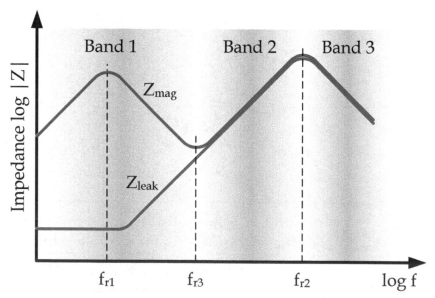

Fig. 12. Simplified representation of VT magnetizing and leakage impedance-frequency characteristics

The developed wideband circuit model of VT is based on the classical circuit model of the magnetic coupling presented in previous subsection. Distributed parasitic capacitances of VT windings are modelled by the lumped capacitances connected only to windings terminals (Fig.13). This assumption reduces noticeable model complexity and allows determining parasitic capacitances based on the measured windings impedances. In the analysed case primary and secondary windings of investigated VT are one side grounded which limits furthermore the number of lumped capacitances necessary to be determined.

Detailed analysis of VT magnetizing and leakage impedance-frequency characteristics and identification of specific resonance frequencies allows estimating parameters of the VT circuit model presented in Fig.13 in the analysed frequency range. The method of determination parasitic lumped capacitances is based on identification of resonance frequencies which are usually possible to determine using the measured impedance

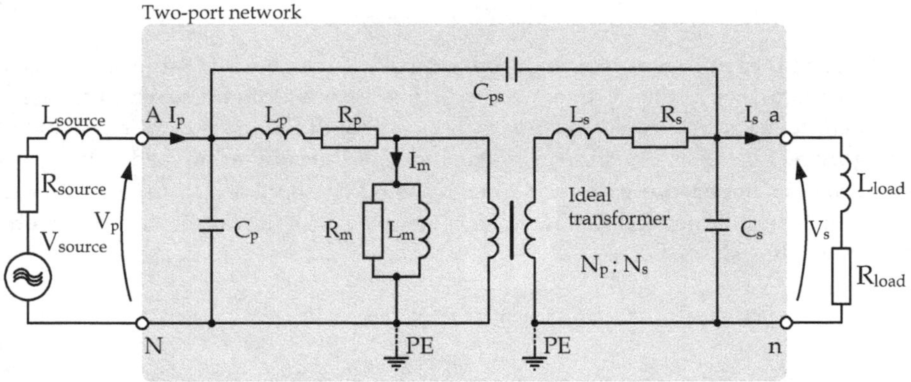

Fig. 13. Broadband circuit model of the VT with lumped parasitic capacitances referenced to windings terminals: $C_p$ – primary winding, $C_s$ – secondary winding, $C_{ps}$ – inter-winding

characteristics (Fig. 10 and 11). Impedance characteristic below resonance frequencies allows to determine adequate inductances and corresponding parasitic capacitances which can be calculated based on the identified resonance frequencies (13), (14). Winding inductances and parasitic capacitances vary slightly with frequency increase therefore matching approximations should be taken into account for simplification.

$$f_{r1} \approx \frac{1}{2\pi\sqrt{L_m\left(C_p + C_s\right)}} \tag{13}$$

$$f_{r2} \approx \frac{1}{2\pi\sqrt{\left(L_p + L_s\right)L_m\left(C_s + C_{ps}\right)}} \tag{14}$$

For analytical evaluation of proposed circuit model two-port network representation can be efficiently used (15) and the final formula (16) for the VT transfer function characteristic vs. frequency can be determined. Transmission matrix representation of the defined two-port network model of VT allows furthermore for consideration of the influence of source impedance (17) and load impedances (18) which have to be considered in broadband analysis.

$$\begin{bmatrix} \underline{V}_p(j\omega) \\ \underline{I}_p(j\omega) \end{bmatrix} = \begin{bmatrix} \underline{A}(j\omega) & \underline{B}(j\omega) \\ \underline{C}(j\omega) & \underline{D}(j\omega) \end{bmatrix} \begin{bmatrix} \underline{V}_s(j\omega) \\ \underline{I}_s(j\omega) \end{bmatrix} \tag{15}$$

$$\underline{H}(j\omega) = \frac{U_s(j\omega)}{U_{source}(j\omega)} = \frac{Z_{load}}{\underline{A}\underline{Z}_{load} + \underline{B} + \underline{C}\underline{Z}_{source}\underline{Z}_{load} + \underline{D}\underline{Z}_{source}} \tag{16}$$

$$\underline{Z}_{source} = R_{source} + j\omega L_{source} \tag{17}$$

$$\underline{Z}_{load} = R_{load} + j\omega L_{load} \tag{18}$$

The $A,B,\ C$ and $D$ coefficients of transmission matrix $T$ (19) can be used for fully charactering electrical performance of the VT in any external condition and its frequency characteristics can be defined by measurement for a specific load condition. Based on the relationship resulting from (15); $A(j\omega)$ is a complex voltage transfer function at no load (20), $B(j\omega)$ is a complex transfer impedance with the secondary winding shorted (21), $C(j\omega)$ is a complex transfer admittance at no load (22), $D(j\omega)$ is a complex current transfer function with the secondary winding shorted (23).

$$T = \begin{bmatrix} \underline{A}(j\omega) & \underline{B}(j\omega) \\ \underline{C}(j\omega) & \underline{D}(j\omega) \end{bmatrix} \tag{19}$$

$$A(j\omega) = \frac{\underline{V}_p(j\omega)}{\underline{V}_s(j\omega)}\bigg|_{\underline{I}_s=0} \tag{20}$$

$$B(j\omega) = \frac{\underline{V}_p(j\omega)}{\underline{I}_s(j\omega)}\bigg|_{\underline{V}_s=0} \tag{21}$$

$$C(j\omega) = \frac{\underline{I}_p(j\omega)}{\underline{V}_s(j\omega)}\bigg|_{\underline{I}_s=0} \tag{22}$$

$$D(j\omega) = \frac{\underline{I}_p(j\omega)}{\underline{I}_s(j\omega)}\bigg|_{\underline{V}_s=0} \tag{23}$$

## 5. Simulation investigation of voltage transformer circuit model

The investigated circuit model of VT can be examined by simulation in any PSpice compatible environment in the conducted disturbance propagation frequency range up to $30\,MHz$. The essential verification of VT model adequacy has been done by determining magnetizing and leakage impedance characteristics which allows verifying model representation adequacy of magnetic coupling between windings. Obtained impedance characteristics are presented in Fig.14 and should be compared to the corresponding measurement results presented in Fig.10 and Fig.11.

Obtained simulation results confirm generally broad-spectrum impedance variations within few $dB$ accuracy margin but many greater than few $dB$ discrepancies can be observed especially close to the resonance frequencies and in frequency range above a few kHz. It should be underlined that the expected compliance level cannot be too high because of many simplifications implemented in the model. The most significant limitation of the evaluated circuit model is associated with lumped representation of winding parasitic capacitances with relation to windings terminals only.

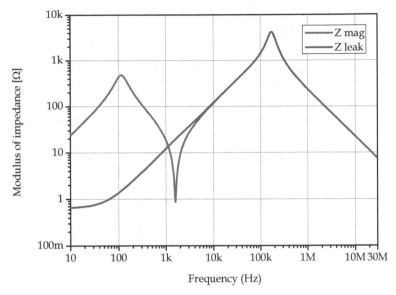

Fig. 14. Magnetizing and leakage impedances of the evaluated VT calculated using developed circuit model

Despite the relatively low accuracy, the developed VT circuit model can be used for simulation analysis of the influence of the VT parameters and its load on the voltage transfer ratio frequency characteristic. The exemplary simulation results of VT voltage transfer ratio characteristics calculated for different resistive loads are presented in Fig. 15.

Based on the presented simulation results it can be noticed that the VT voltage transfer characteristic change essentially for frequencies higher than the main resonance frequency observed on the leakage impedance, which is about 100 kHz for the evaluated case. Above this frequency VT voltage transfer ratio depends mainly on winding parasitic capacitances and magnetic coupling between windings becames less meaningful.

Simulation results demonstrate that in frequency range close to leakage impedance resonance VT load has the major influence on the VT transfer characteristic. Increase of resistive VT load reduces significantly VT voltage transfer ratio around this frequency. Obtained simulation results confirm that according to the analytical investigation Eq. (8), VT load rate has significant influence on VT performance in high frequency range and limits usually its pass band. According to the presented simulated transfer characteristics it is possible to expand the pass band width by lowering the VT load. Unfortunately, low VT load can intensify adverse effects of any resonances which might arise in VT.

VT load character, capacitive or inductive, has also essential influence on the voltage transfer ratio frequency characteristics. Simulation results for VT loaded with the same impedance determined for power frequency and different power factor (1 resistive, 0.7 inductive and 0.7 capacitive) is presented in Fig. 16. The obtained results demonstrate positive impact of inductive character of VT load on the pass band width. Inductive VT load broadens the pass band towards the resonance frequency and causes effects of resonance to be more sharp. In the evaluated case, the change of the VT load from purely resistive to inductive (PF=0.7 @ 50 Hz) increases the 3 dB cut-off frequency few times.

Capacitive character of VT load induces opposite effect and narrows VT pass band width significantly. Similarly, the change of VT load from purely resistive to capacitive (*PF=-0.7 @ 50 Hz*) decrease of VT pass band width is about tens times.

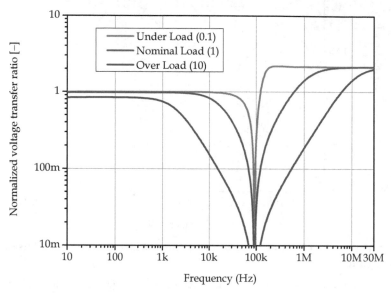

Fig. 15. Simulation results of the influence of resistive load of VT on voltage transfer ratio frequency characteristic

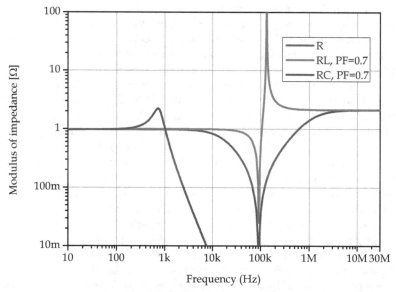

Fig. 16. Influence of the character of VT load on voltage transfer ratio frequency characteristic – simulation results

## 6. Experimental tests of voltage transformer transfer characteristic

Experimental investigations have been done for voltage transformers typically used in MV power system with primary and secondary windings grounded. Exemplary measurement results presented in this chapter have been obtained for VT of *50 VA* rated power and *20 kV/0.1 kV* nominal transformation ratio. Parameters of the proposed VT circuit model for simulation have been identified by analysis of secondary windings impedance-frequency characteristics measured for no load condition (magnetizing inductance – Fig.10) and short circuit condition (leakage inductance – Fig.11). Measurements have been done in frequency range from *10 Hz* up to *30 MHz*, which is a range typically used for the analysis of conducted disturbances in power system. Particular attention has been paid to the frequency range below *10 kHz* which is obligatory for power quality analysis, especially for analysis of power system voltage harmonics related phenomena.

The measured voltage transfer characteristics of evaluated VT for nominal load and no load conditions are presented in Fig. 17. Based on these results the *3 dB* high frequency pass band of the evaluated VT can be estimated to be about *2 kHz*. For frequencies higher than 2 kHz a number of less meaningful resonances are clearly visible on the VT voltage transfer characteristic which are not adequately characterized by the evaluated circuit model. This inadequacy is associated with extra internal resonances appearing in windings which cannot be properly represented by lumped parasitic capacitances referenced only to windings terminals. In order to model these phenomena more complex circuit model are required which take into account more detailed distributed representation of partial parasitic capacitance of VT primary winding.

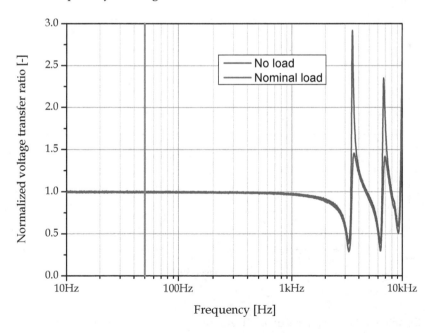

Fig. 17. Normalized voltage transfer ratio of the VT for no load and nominal resistive load

Comparison of voltage transfer characteristic measured for no load and nominal resistive load condition confirms that the influence of the level of the resistive load is mostly observable for frequencies close to the resonance frequencies. For these frequency ranges the voltage transfer ratio can vary even few times due to the VT load change.

Comparison of VT leakage and magnetizing impedances allow for preliminary approximation of the VT pass band cut-off frequency. In Fig. 18 correlation between VT impedances and the measured voltage transfer ratio is presented. Based on this comparison it can be noticed that:

- firstly, for the frequency range where magnetizing inductance is evidently higher than leakage impedance (Band 1 according to Fig. 12) the magnetic coupling between VT windings is tough, the VT voltage transfer characteristic is nearly flat and relatively weakly dependent on load,
- secondly, for the frequency range where magnetizing and leakage inductances are comparable (Band 2 according to Fig. 12) the VT voltage transfer characteristic is hardly dependent of VT load character and the influence of internal distribution of parasitic capacitances of winding is manifested by extra local parasitic resonance occurrence.

Additional effects of parasitic capacitance distribution, which are not sufficiently represented by evaluated simplified circuit model, justify narrower pass band of VT obtained by experimental investigation (about *2 kHz*) with comparison to simulation results (about *20 kHz*).

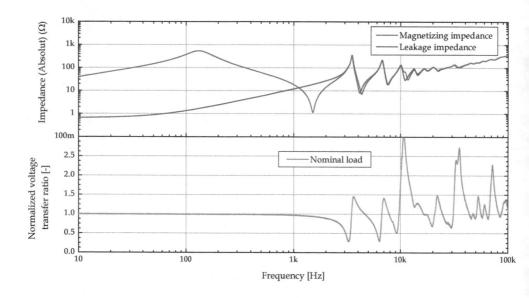

Fig. 18. Correlation between measured VT voltage transfer characteristic and magnetizing and leakage impedance-frequency characteristics

Accurateness of magnitude and phase voltage transfer characteristics of VT is a fundamental aspect for identification and measurement of power quality related phenomena in power system. For the investigated VT the voltage transfer ratio and voltage phase shift characteristics have been measured to reveal measurement accuracy problems of power quality assessment in MV systems. Magnitudes versus phase transfer characteristic of VT measured for different frequency ranges typically used in power quality measurement systems (up to 40th harmonic and up to 9 kHz) are presented in Fig. 19 and Fig. 20. Experimental investigations prove that magnitude and phase errors increase noticeably with frequency. In frequency range up to 2 kHz, the highest magnitude error of about 11 % and phase shift error almost 8°, have been obtained for frequency 2 kHz. These results confirm that voltage harmonics measurement in MV grids by using VT can be not accurate enough in applications with noticeable harmonic content above approximately 1 kHz.

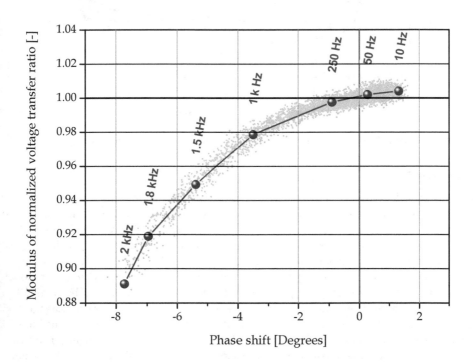

Fig. 19. Normalized VT voltage ratio vs. phase shift angle for frequency band up to 2 kHz

Magnitudes and phase inaccuracy of VT obtained in frequency range from 2 kHz up to 9 kHz (Fig. 20) are evidently greater and its frequency dependence is more complex, therefore more difficult to model using simplified circuit models. Magnitude errors in this frequency range reach almost 180% and phase shift error almost 80°, which cannot be accepted in power quality measurement applications.

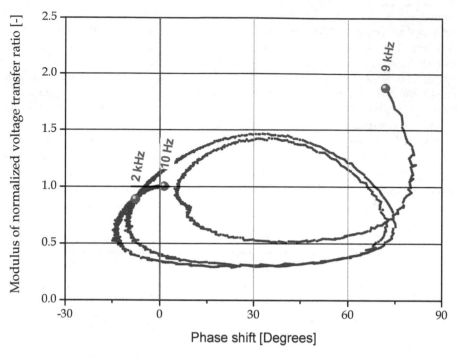

Fig. 20. Normalized VT voltage ratio vs. phase shift angle for frequency band up to 9 kHz

## 6. Conclusions

Modelling of VT voltage transfer characteristic in wide frequency range is rather challenging. Main problems with accurate modelling using circuit models are related to windings' parasitic capacitances and especially identification of its unequal distribution along windings. To model the influence of parasitic capacitive couplings existing in a typical VT several simplifications should be considered. The method of VT parasitic capacitances analysis based on the lumped representation is often used and particularly rational, nevertheless limits the frequency range within which acceptable accuracy can be obtained.

Its parameters can be determined based only on wideband measurement of leakage and magnetizing impedances, unfortunately it can be successfully used only in the limited frequency range. For typical VT used in MV grids the flatten fragment of transfer characteristic can be obtained usually only up to few kHz. Above this frequency VT usually exhibit a number of resonances which change evidently its transfer characteristic and cannot be reflected adequately by simplified circuit models. Wideband performance of VT in a particular application is also noticeably related to its load level and character (inductive or capacitive). For typical VT it is possible to improve slightly its wideband performance by lowering its load level or by changing its character into inductive, but it usually requires laborious experimental verification. Despite of recognized restrictions and limited accuracy of the developed circuit model it can be successfully used for approximate assessment of VT pass band.

The use of VT in power quality monitoring MV grids influence essentially finally obtained measurement accuracy. In power quality measurement applications where dominating harmonics emission is expected only in frequency range below 2 kHz VTs can provide sufficient accuracy in many applications, nevertheless its voltage transfer characteristic should be carefully verified with taking into account particular operating conditions. Nowadays, much wider than up to 2 kHz harmonics emission spectrum can be injected into the power system, especially by contemporary high power electronic applications. In this frequency range from 2 kHz up to 9 kHz, which is already well specified by harmonic emission limitation standards, typically used VT are not reliable enough. Measurement errors in frequency range up to 9 kHz are usually not acceptable, because of resonance effects which commonly appear and are difficult to predict.

## 7. References

Islam, S.M.; Coates, K.M.; Ledwich, G.; Identification of high frequency transformer equivalent circuit using Matlab from frequency domain data. Thirty-Second IAS Annual Meeting, IAS '97., Conference Record of the 1997 IEEE Industry Applications Conference, 1997

Kadar, L.; Hacksel, P.; Wikston, J.; The effect of current and voltage transformers accuracy on harmonic measurements in electric arc furnaces., IEEE Transactions on Industry Applications, Volume 33, Issue 3, May-June 1997 Page(s):780 - 783

Klatt, M.; Meyer, J.; Elst, M.; Schegner, P.; Frequency Responses of MV voltage transformers in the range of 50 Hz to 10 kHz. 14th International Conference on Harmonics and Quality of Power (ICHQP), 2010

Łuszcz J.; Conducted EMI propagation modelling in the wound components. Seventeenth International Wrocław Symposium and Exhibition on Electromagnetic Compatibility, EMC 2004

Łuszcz J.; Iron Core Inductor High Frequency Circuit Model for EMC Application. Coil Winding International & Electrical Insulation Magazine. Volume 28, Issue 1, 2004

Mahesh, G.; George, B.; Jayashankar, V.; Kumar, V.J.; Instrument transformer performance under distorted-conditions. India Annual Conference, 2004. Proceedings of the IEEE INDICON 2004. Page(s):468 – 471

Mohamed, R.; Markovsky, I.; Lewin, P.L.; Modeling and Parameter Estimation of High Voltage Transformer Using Rational Transfer Function State Space Approach. Annual Report Conference on Electrical Insulation and Dielectric Phenomena, 2008. CEIDP 2008

Seljeseth, H.; Saethre, E.A.; Ohnstad, T.; Lien, I.; Voltage transformer frequency response. Measuring harmonics in Norwegian 300 kV and 132 kV power systems., 1998. Proceedings. 8th International Conference on Harmonics And Quality of Power. Volume 2, 14-16 Oct. 1998 Page(s):820 - 824 vol.2

Shibuya, Y.; Fujita, S.; High frequency model and transient response of transformer windings, Transmission and Distribution Conference 2002: 6-10 Oct. 2002

Vermeulen, H.J.; Dann, L.R.; van Rooijen, J.; Equivalent circuit modelling of a capacitive voltage transformer for power system harmonic frequencies, IEEE Transactions on Power Delivery, Volume 10, Issue 4, Oct. 1995

Yao Xiao; Jun Fu; Bin Hu; Xiaoping Li; Chunnian Deng; Problems of voltage transducer in
    harmonic measurement., IEEE Transactions on Power Delivery, Volume 19, Issue 3,
    July 2004 Page(s):1483 – 1487

# 3

# On the Reliability of Real Measurement Data for Assessing Power Quality Disturbances

Alexandre Brandao Nassif
*Hydro One Inc.,Toronto, ON,*
*Canada*

## 1. Introduction

Power quality assessment is a power engineering field that is first and foremost driven by real data measurements. All the power quality assessment applications rely on results from real data processing. Take as an example the art of harmonic filter design, which is an engineering field notoriously known for relying on simulation-based planning; in this technical assessment, data recordings are indirectly used for finding the frequency response (or $R$-$X$ plots) of the system impedance that is/are in turn used to determine the filters' tuning frequencies (Kimbark, 1971).

With so much reliance on the acquired data, the quality of such has become a very sensitive issue in power quality. An imperative action is to always employ high-resolution recording equipment in any instance of power quality analysis. Nevertheless, high-resolution equipment does not guarantee data usefulness because the measured data may be inherently of very low energy in a variety of ways. Therefore, to investigate such cases and to propose methods to identify useful data were the motivations for this research. This chapter proposes methods for data selection to be used in two applications where the reliability issue is crucial: the power system impedance estimation and the interharmonic source determination.

### 1.1 The network harmonic impedance estimation

Network impedance is power system parameter of great importance, and its accurate estimation is essential for power system analysis at fundamental and harmonic frequencies. This parameter is deemed of being of great importance for a variety of power system applications, such as evaluating the system short-circuit capacity, or defining the customer harmonic limits (Kimbark, 1971)-(IEEE Std. 519-1992). Several methods have been proposed to measure the network harmonic impedance and are available in literature. In this chapter, the transient-based approach is used to demonstrate the data selection methods. In the transient-based approach, the network impedance is conventionally calculated by using (Robert & Deflandre, 1997)

$$Z_{eq}(h) = -\Delta V(h)/\Delta I(h), \tag{1}$$

where $\Delta V(h)$ and $\Delta I(h)$ are the subtraction in frequency domain of one or more cycles previous to the transient occurrence from the corresponding cycles containing the transient disturbance. The objective of this chapter is not to promote the use of the transient-based approach for determining the network harmonic impedance, nor is it to explain the method in detail. The reader is encouraged to consult (Robert & Deflandre, 1997) for details. In this application, the level of accuracy of such estimation can be supported by a set of indices, which are (but not limited to) the quantization noise in the data acquisition, the frequency resolution, the energy levels, and the scattering of the results obtained from the data.

## 1.2 The Interharmonics measurement

Interharmonics are spectral components which frequencies are non-integer multiples of the supply fundamental frequency. This power quality event represents the target of the second application of the proposed reliability criteria. Diagnosing interharmonic problems is a difficult task for a number of reasons: (1) interharmonics do not manifest themselves in known and/or fixed frequencies, as they vary with the operating conditions of the interharmonic-producing load; (2) interharmonics can cause flicker in addition to distorting the waveforms, which makes them more harmful than harmonics; (3) they are hard to analyze, as they are related to the problem of waveform modulation (IEEE Task Force, 2007). The most common effects of interharmonics have been well documented in literature (IEEE Task Force, 2007), (Ghartemani & Iravani, 2005)-(IEEE Interhamonic Task Force, 1997), (Yacamini, 1996). Much of the published material on interharmonics has identified the importance of determining the interharmonic source (Nassif et al, 2009, 2010a, 2010b). Only after the interharmonic source is identified, it is possible to assess the rate of responsibility and take suitable measures to design mitigation schemes. Interharmonic current spectral bins, which are typically of very low magnitude, are prone to suffer from their inherently low energy level. Due to this difficulty, the motivation of the proposed reliability criteria is to strengthen existing methods for determining the source of interharmonics and flicker which rely on the active power index (Kim et al, 2005), (Axelberg et al, 2008).

## 1.3 Objectives and outline

The objective of this research is to present a set of reliability criteria to evaluate recorded data used to assess power quality disturbances. The targets of the proposed methods are the data used in the determination of the network harmonic impedance and the identification of interharmonic sources. This chapter is structured as follows. Section 2 presents the data reliability criteria to be applied to both challenges. Section 3 presents the harmonic impedance determination problem and section 4 presents a network determination case study. Section 5 presents the interharmonic source determination problem and sections 6 and 7 present two case studies. Section 8 presents general conclusions and recommendations.

## 2. Data reliability criteria

This section is intended to present the main data reliability criteria proposed to be employed in the power quality applications addressed in this chapter. The criteria are applied in a slightly different manner to fit the nature of each problem. As it will be explained in this chapter, in the context of the network impedance estimation, the concern is $\Delta I(f)$ and $\Delta V(f)$ (the variation of the voltage or current), whereas for the case of interharmonic measurement, the concern is the value of $I(f)$ and $V(f)$. The reason for this will be explained in more detail in sections 3 and 4, and at this point it is just important to keep in mind that the introduced criteria is applied in both cases, but with this slight difference.

## 2.1 The energy level index

As shown in (1), the network impedance determination is heavily reliant on $\Delta I(f)$, which is the denominator of the expression. Any inaccuracy on this parameter can result in great numerical deviance of the harmonic impedance accurate estimation. Therefore, the $\Delta I(f)$ energy level is of great concern. For this application, a threshold was suggested in (Xu et al, 2002) and is present in (2). If the calculated index is lower than the threshold level, the results obtained using these values are considered unreliable.

$$I_{threshold} = \frac{\Delta I(f)}{\Delta I(60Hz)} > 1\%.$$ (2)

Fig. 1 shows an example on how this criterion can be used. The energy level for $\Delta I(f)$ is compared with the threshold. For this case, frequencies around the 25th harmonic order (1500Hz) are unreliable according to this criterion.

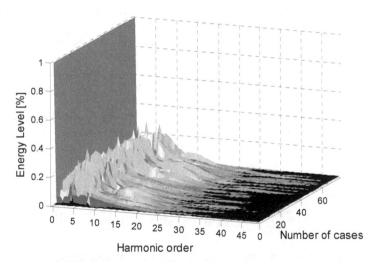

Fig. 1. Energy level of $\Delta I(f)$ seen in a three-dimensional plot

## 2.2 Frequency-domain coherence index

This index is used in the problem of the network impedance estimation, which relies on the transient portion of the recorded voltages and currents (section 3 presents the method in detail). The random nature of a transient makes it a suitable application for using the power density spectrum (Morched & Kundur, 1987). The autocorrelation function of a random process is the appropriate statistical average, and the Fourier transform of the autocorrelation function provides the transformation from time domain to frequency domain, resulting in the power density spectrum.

This relationship can be understood as a transfer function. The concept of transfer function using the power spectral method based on correlation functions can be treated as the result from dividing the cross-power spectrum by the auto-power spectrum. For electrical power systems, if the output is the voltage and the input is the current, the transfer function is the impedance response of the system (Morched & Kundur, 1987). The degree of accuracy of the transfer function estimation can be assessed by the coherence function, which gives a measure of the power in the system output due to the input. This index is used as a data selection/rejection criterion and is given by

$$\gamma_{VI}(f) = \frac{|P_{VI}(f)|^2}{P_{VV}(f)P_{II}(f)},$$

(3)

where $P_{VI}(f)$ is the cross-power spectrum of the voltage and current, which is obtained by the Fourier transform of the correlation between the two signals. Similarly, the auto-power spectrum $P_{VV}(f)$ and $P_{II}(f)$ are the Fourier transforms of the voltage and current auto-correlation, respectively. By using the coherence function, it is typically revealed that a great deal of data falls within the category where input and output do not constitute a cause-effect relationship, which is the primary requirement of a transfer function.

### 2.3 Time-domain correlation between interharmonic current and voltage spectra

This index is used for the interharmonic source detection analysis, and is the time-domain twofold of the coherence index used for the harmonic system impedance. The criterion is supported by the fact that, if genuine interharmonics do exist, voltage and current spectra should show a correlation (Li et al, 2001) because an interharmonic injection will result in a voltage across the system impedance, and therefore both the voltage and current should show similar trends at that frequency. As many measurement snapshots are taken, the variation over time of the interharmonic voltage and current trends are observed, and their correlation is analyzed. In order to quantify this similarity, the correlation coefficient is used (Harnett, 1982):

$$r(ih) = \frac{n\sum_{i=1}^{n} I_{IH}(i)V_{IH}(i) - \sum_{i=1}^{n} I_{IH}(i)\sum_{i=1}^{n} V_{IH}(i)}{\sqrt{\left[n\sum_{i=1}^{n} I_{IH}^2(i) - \left(\sum_{i=1}^{n} I_{IH}(i)\right)^2\right]\left[n\sum_{i=1}^{n} V_{IH}^2(i) - \left(\sum_{i=1}^{n} V_{IH}(i)\right)^2\right]}},$$

(4)

where $I_{IH}$ and $V_{IH}$ are the interharmonic frequency current and the voltage magnitudes of the $n$-snapshot interharmonic data, respectively. Frequencies showing the calculated correlation coefficient lower than an established threshold should not be reliable, as they may not be genuine interharmonics (Li et al, 2001).

### 2.4 Statistical data filtering and confidence intervals

In many power quality applications, the measured data are used in calculations to obtain parameters that are subsequently used in further analyses. For example, in the network impedance estimation problem, the calculated resistance of the network may vary from 0.0060 to 0.0905 (ohms) in different snapshots (see Fig. 2). The resistance of the associated network is the average of these results. Most of the calculated resistances are between 0.0654 and 0.0905 (ohms). Those values that are numerically distant from the rest of the data (shown inside the circles) may spoil the final result as those data are probably gross results. As per statistics theory, in the case of normally distributed data, 97 percent of the observations will differ by less than three times the standard deviation [14]. In the study presented in this chapter, the three standard deviation criterion is utilized to statistically filter the outlier data.

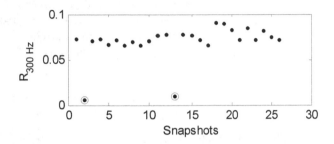

Fig. 2. Calculated 5th harmonic resistance over a number of snapshots

In the example presented in Fig. 2, once the resistance of the network is achieved by averaging the filtered data, the confidence on the obtained results might be questioned. Instead of estimating the parameter by a single value, an interval likely to include the parameter is evaluated. Confidence intervals are used to indicate the reliability of such an estimate (Harnett, 1982). How likely the interval is to contain the parameter is determined by the confidence level or confidence coefficient. Increasing the desired confidence level will widen the confidence interval. For example, a 90% confidence interval for the achieved resistance will result in a 0.0717 ± 0.0055 confidence interval. In the other words, the resistance of the network is likely to be between 0.662 and 0.772 (ohms) with a probability of 90%.

Fig. 3 shows the calculated harmonic impedance of the network. Error bars are used to show the confidence intervals of the results. Larger confidence intervals present less reliable values. In this regard, the estimated resistance at 420 Hz is more reliable than its counterpart at 300 Hz.

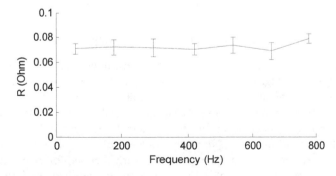

Fig. 3. Selected 5th harmonic resistance data showing confidence intervals.

## 2.5 Quantization error

Quantization refers to the digitalization step of the data acquisition equipment. This value dictates the magnitude threshold that a measurement must have to be free of measurement quantization noise (Oppenheim & Shafer, 1999). The A/D conversion introduces quantization error. The data collected are in the form of digital values while the actual data are in analog form. So the data are digitalized with an A/D converter. The error associated with this conversion is the quantization step. As the energy of current signals drops to a level comparable to that of quantization noises, the signal may be corrupted, and the data will, therefore, be unreliable. For this reason, if the harmonic currents are of magnitude lower than that of the quantization error, they should not be trusted. This criterion was developed as follows:

1. The step size of the quantizer is

$$\Delta = V_{in}/2^n ,$$                                                                        (5)

where $n$ is the number of bits and $V_{in}$ is the input range.
2. The current probe ratio is $k_{probe}$, which is the ratio V/A.
3. Therefore, the step size in amperes is

$$\Delta_I = \Delta/k_{probe} .$$                                                                (6)

4. Finally, the maximum quantization error will be half of the step size.

The input range, number of bits and current probe ratio will depend on the data acquisition equipment and measurement set up. The measurements presented later in this chapter are acquired by high-resolution equipment (NI-6020E - 100kbps, 12-bit, 8 channels). For the case of the system impedance estimation, equation (7) should hold true in order to generate reliable results for single-phase systems. This criterion is also used for $\Delta V(f)$:

$$|\Delta I(f)| > I_{error} .$$                                                                  (7)

For the interharmonic case, the interharmonic current level $I(ih)$ is monitored rather than the $\Delta I(f)$:

$$|I(ih)| > I_{error} .$$                                                                        (8)

## 3. Network harmonic impedance estimation by using measured data

The problem of the network harmonic impedance estimation by using measured data is explained in this section. Fig. 4 presents a typical scenario where measurements are taken to estimate the system harmonic impedance. Voltage and current probes are installed at the interface point between the network and the customer, called the point of common coupling (PCC). These probes are connected to the national instrument NI-6020E 12-bit data acquisition system with a 100 kHz sampling rate controlled by a laptop computer. Using this data-acquisition system, 256 samples per cycle were obtained for each waveform. In Fig. 4, the impedance $Z_{eq}$ is the equivalent impedance of the transmission and distribution lines, and of the step-down and step-up transformers.

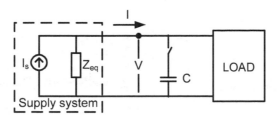

Fig. 4. Equivalent circuit for system impedance measurement.

Many methods that deal with measuring the harmonic impedance have been proposed and published (Xu et al, 2002), (Morched & Kundur, 1987), (Oliveira et al, 1991). They can be classified as either invasive or non-invasive methods. Invasive methods are intended to produce a disturbance with energy high enough to change the state of the system to a different post-disturbance state. Such change in the system is necessary in order to obtain data records to satisfy (9) and (10), but low enough not to affect the operation of network equipment. The applied disturbance in the system generally causes an obvious transient in the voltage and current waveforms. The transient voltage and current data are used to obtain the impedance at harmonic frequencies. For the case presented in this chapter, the source of disturbance is a low voltage capacitor bank, but other devices can also be used, as explained in (Xu et al, 2002), (Morched & Kundur, 1987), (Oliveira et al, 1991)

Therefore, the transient signal is extracted by subtracting one or more intact pre-disturbance cycles from the cycles containing the transient, as

$$V_{transient} = V_{disturbance} - V_{pre\_disturbance} = \Delta V,$$
$$I_{transient} = I_{disturbance} - I_{pre\_disturbance} = \Delta I. \tag{9}$$

Finally, the network impedance is calculated by using

$$Z_{eq}(h) = -\Delta V(h)/\Delta I(h). \tag{10}$$

### 3.1 Characterization of the capacitor switching transient

Traditionally, transients are characterized by their magnitude and duration. For the application of network impedance estimation, the harmonic content of a transient is a very useful piece of information. A transient due to the switching of a capacitor has the following characteristics (IEEE Std. 1159-1995):

Magnitude: up to 2 times the pre-existing voltage (assuming a previously discharged capacitor).
• Duration: From 0.3ms to 50ms.
• Main frequency component: 300Hz to 5 kHz.

The energization of the capacitor bank (isolated switching) typically results in a medium-frequency oscillatory voltage transient with a primary frequency between 300 and 900 Hz and magnitude of 1.3-1.5 p.u., and not longer than two 60Hz cycles. Fig. 5 shows typical transient waveforms and frequency contents due to a capacitor switching. For this case, the higher frequency components (except the fundamental component) are around 5th to 10th harmonic (300-600Hz).

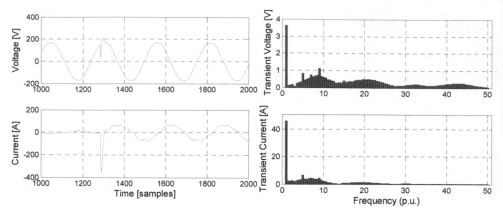

Fig. 5. Characterization of the transients resulting from a capacitor switching: (a) voltage and current waveforms during a disturbance, (b) Transient waveforms and frequency contents.

## 3.2 Transient Identification
The perfect extraction of the transient is needed for the present application. Several classification methods were proposed to address this problem, such as neural networks and wavelet transforms (Anis & Morcos, 2002). Some other methods use criteria detection based on absolute peak magnitude, the principal frequency and the event duration less than 1 cycle (Sabin et al, 1999).

In this chapter, a simple approach is proposed to perform this task. It calculates the numerical derivative of the time-domain signals, and assumes that if a transient occurred, this derivative should be higher than 10. As a result, the numerical algorithm monitors the recorded waveforms and calculates the derivatives at each data sample; when this derivative is higher than 10, it can be concluded that a capacitor switching occurred.

## 4. Impedance measurement case study

More than 120 field tests have been carried out in most of the major utilities in Canada (in the provinces of British Columbia, Ontario, Alberta, Quebec, Nova Scotia and Manitoba), and a representative case is presented in this section. Over 70 snapshots (capacitor switching events) were taken at this site. Using the techniques described in section II, the impedance results were obtained and are presented in Fig. 6. This figure shows that in the range of 1200-1750Hz there is an unexpected behavior in both components of the impedance. A resonant condition may be the reason of this sudden change. However, it might be caused by unreliable data instead. Further investigation is needed in order to provide a conclusion for this case.

Based on extensive experience acquired by dealing with the collected data, the following thresholds were proposed for each index:

- Energy level: $\Delta I(f) > 1\%$ and $\Delta V(f) > 1\%$.
- Coherence: $\gamma(f) \geq 0.95$.
- Standard deviation: $\delta \geq 0.5$.
- Quantization error: $\Delta I(f) > 0.0244A$.

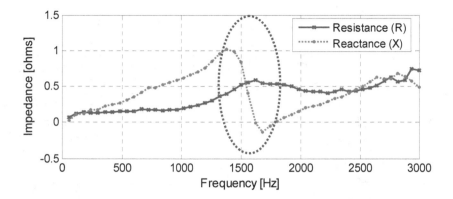

Fig. 6. Harmonic impedance for a sample field test used in the case study.

As the reliability criteria are applied, it is useful to define the following ratio of success:

$$Ratio = Successfull\ cases/Total\ cases \times 100\%, \tag{11}$$

where each case is one data snapshot taken at each site.
Fig. 7 shows the success rate of cases for each index in function of frequency. Fig. 7a. also shows that the application of this index will affect $\Delta I$ much more than for $\Delta V$, since the latter is acquired using voltage probes, which are inherently much more reliable. Since the impedance measurement is calculated from the ratio $-\Delta V/\Delta I$, the voltage threshold is applied to the denominator and is therefore less sensitive, as shown in (Xu et al, 2002). Fig. 7b. shows that the coherence index does not reveal much information about reliability of the measurements; however it provides an indicator of the principal frequency of the transient signals, highlighted in the dotted circle. The standard deviation results presented in Fig. 7c. show that the impedances measured at frequencies between 1260 and 2000 Hz are very spread out and are, therefore, unreliable. The same situation occurs for frequencies above 2610 Hz. These results agree with those presented in Fig. 7a. for the threshold used for $\Delta I$. Fig. 7d. shows that the quantization is not a critical issue and the measurements taken in the field are accurate enough to overcome quantization noises. However the low quantization values, especially for current, are of lower values for the unreliable ranges presented in Fig. 7a. and Fig. 7c.

## 5. Interharmonic source determination

In harmonic analysis, many polluters are usually present in a power distribution system for each harmonic order because power system harmonics always occur in fixed frequencies, i.e., integer multiples of the fundamental frequency. All harmonic loads usually generate all harmonic orders, and therefore, it is common to try to determine the harmonic contribution of each load rather than the harmonic sources. As opposed to harmonics, interharmonics are almost always generated by a single polluter. This property of interharmonics can be explained as follows.

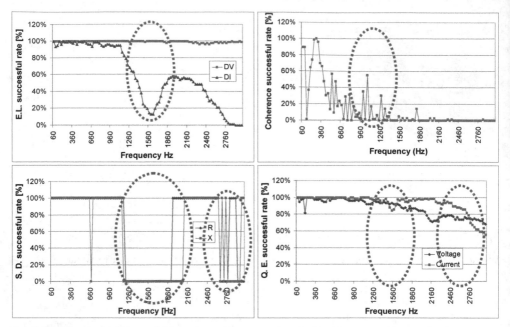

Fig. 7. Indices in function of frequency: (a) energy level, (b) coherence, (c) standard deviation, (d) quantization error.

The main interharmonic sources are adjustable speed drives (ASDs) with a $p_1$-pulse rectifier and a $p_2$-pulse inverter and periodically varying loads such as arc furnaces. Their interharmonic generation characteristics can be expressed as in (12) for ASDs (Yacamini, 1996) and (13) for periodically varying loads (IEEE Task Force, 2007), respectively:

$$f_{IH} = \left| (p_1 m \pm 1) f \pm p_2 n f_z \right|, \quad m = 0,1,2...; \quad n = 1,2,3..., \tag{12}$$

where $f$ and $f_z$ are the fundamental and drive-operating frequency.

$$f_{IH} = \left| f \pm n f_v \right|, \quad n = 1,2,3..., \tag{13}$$

where $f_v$ is the load-varying frequency. According to equations (12) and (13), the interharmonic frequency depends on many factors such as the number of pulses of the converter and inverter, the drive-operating frequency, or the load-varying frequency. Therefore, the same frequency of interharmonics is rarely generated by more than one customer.

Based on the above analysis, for interharmonic source determination, the analysis can be limited to the case of a single source for each of the interharmonic components. The most popular method currently being used to identify the interharmonic sources is based on the active power index. Fig. 8 helps to explain the power direction method. For this problem, the polluter side is usually assumed to be represented by its respective Norton equivalent circuit. Fig. 8 shows two different scenarios at the metering point between the upstream (system) and the downstream (customer) sides. Fig. 8a and Fig. 8b show the case where the interharmonic components come from the upstream side and the downstream side, respectively. The circuits presented in Fig. 8 are used for each frequency.

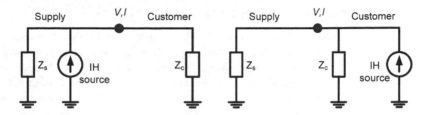

a. IH comes from Supply side        b. IH comes from Customer side

Fig. 8. Determination of interharmonic source – two different scenarios.

The interharmonic active power can be obtained from the voltage and current measurements at a metering point as

$$P_{IH} = \text{Re}\{V_{IH} \times I_{IH}^{*}\} = |V_{IH}||I_{IH}|\cos(\phi_{IH}),$$    (14)

where $|V_{IH}|$ and $|I_{IH}|$ are the interharmonic voltage and current magnitudes, respectively, and $\varphi_{IH}$ is the angle displacement between the interharmonic voltage and current.

The conclusion of the power direction method, therefore, is the following (Kim et al, 2005), (Axelberg, 2008):

• If $P_{IH} > 0$, the interharmonic component comes from the upstream side.
• If $P_{IH} < 0$, the interharmonic component comes from the downstream side.

If this criterion is extended to a multi-feeding system like that shown in Fig. 9, the interharmonic source for each interharmonic can be identified. In such a case, monitoring equipment should be placed at each feeder suspected of injecting interharmonics into the system. For the system side measurements (point A), if the measured $P_{IH} > 0$, the interharmonic component comes from the system. For the customer side measurements (points B and C), if the measured $P_{IH} < 0$, the interharmonic component comes from the measured customer.

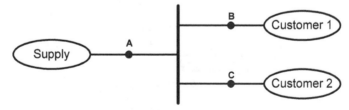

Fig. 9. System diagram for locating the interharmonic source.

As a given interharmonic frequency has only one source, the power direction method (in theory) could always reveal the interharmonic source correctly. In reality, as the active power of the interharmonics is typically very small, the measured data may not be reliable, so this index may not provide reliable conclusions. On the other hand, the angle $\varphi_{IH}$ can be very close to either $\pi/2$ or $-\pi/2$, oscillating around these angles because of measurement errors, and resulting in the measured active power index swinging its sign and potentially causing misjudgment. The drifting nature of interharmonics in frequency and the supply fundamental frequency variation also influence this inaccuracy.

## 6. Interharmonic source determination case study #1

The case study presented in this section illustrates the worst known effect of the interharmonics at present, which is the interference with control signals at the power line. The Automatic Meter Reading devices (AMR) of a utility company of the province of Alberta had experienced difficulties to receive inbound signals in a large oilfield extraction area. Utility engineers suspected that this interference could be caused by the presence of interharmonics. Field measurements were carried out at the substation feeder and at large customers that were suspected to be interharmonic polluters. The arrangement for the field measurement is shown in Fig. 10. The measured feeder supplies three customers, codenamed Customer 1, Customer 2 and Customer 3. These customers operate large oil extracting drives. The measurements were done through potential transformers and current transformers (PTs and CTs). The data were acquired for a period of two days, taking automatic snapshots of 5 seconds at every minute. The hardware utilized was a National Instruments NI-DAQ6020E, which operates at 100kb/s and has 8-channel capability. With this sampling rate, the recorded waveforms contained 256 points per cycle.

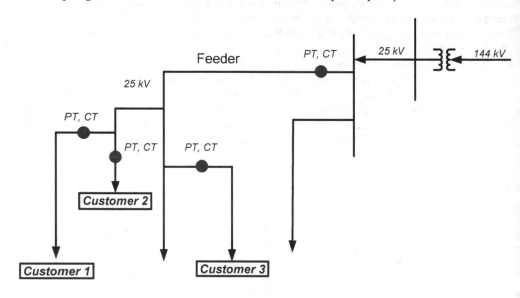

Fig. 10. Field measurement locations at the measured area

After processing all data snapshots taken at the four locations, a spectrum contour plot measured at the feeder is drawn in order to obtain the frequencies of the interharmonic components that are present in this system. Fig. 11 shows the contour plot of the data recorded at the feeder during one of the measured days. From this figure, it can be seen that there are four dominant interharmonic components, which seem to be two pairs: at around 228 Hz and 348 Hz, and 264 Hz and 384 Hz. These components drift a little in frequency due to the change of the drive operation conditions, but they exist inside a narrow frequency range.

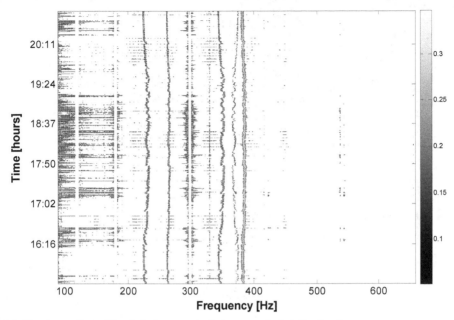

Fig. 11. Contour plot of the interharmonic data recorded at the feeder

The active power index was monitored at the three loads. This is shown in Fig. 12-Fig. 15. The system was observed to be fairly balanced, and therefore only the power in phase A is shown. By looking into these figures, one would conclude that Customer 2 is the source of interharmonics 228Hz, 348Hz and 384Hz, whereas Customer 3 is the source of interharmonics 264Hz and 348Hz. As explained in equation (2), it is almost impossible that an interharmonic component is generated by two sources at the same time. Furthermore, after deeper investigation, it is shown that this apparent identification of the interharmonic polluters is incorrect, and the reliability criteria proposed in this chapter is useful in aiding the researcher to drawing correct conclusions.

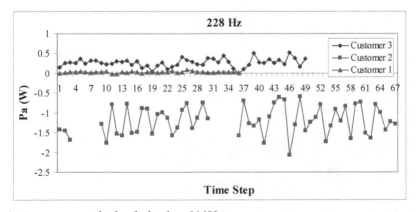

Fig. 12. Active power at the loads for $f_{IH} = 228Hz$

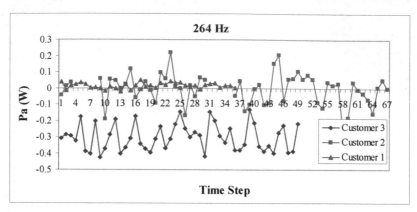

Fig. 13. Active power at the loads for $f_{IH} = 264Hz$

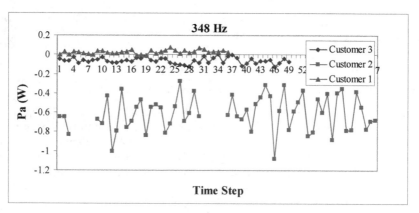

Fig. 14. Active power at the loads for $f_{IH} = 348Hz$

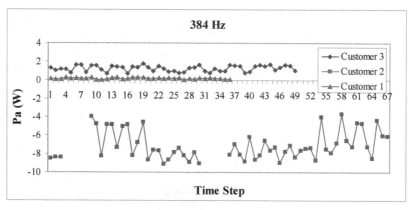

Fig. 15. Active power at the loads for $f_{IH} = 384Hz$

## 6.1 Applying the reliability criteria

The first step to utilize the reliability criteria is to obtain the percentage of snapshots containing measurements with energy levels above the quantization error. This result for the case study is shown in Table 1. According to this criterion, the interharmonic currents measured at the feeder may be unreliable because they are too low as compared to the current fundamental component. This fact does not mean that the measured interharmonics are harmless, but simply that 12 bits of the data acquisition device are not enough to accurately measure their magnitudes. As for the loads, all data are reliable, except those of Customer 3 at 348 Hz.

| IH freq (Hz) | Feeder | Customer 1 | Customer 2 | Customer 3 |
|---|---|---|---|---|
| 228 | 0.00 | 19.44 | 100.00 | 91.84 |
| 264 | 0.00 | 0.00 | 77.19 | 81.63 |
| 348 | 0.00 | 0.00 | 3.51 | 0.00 |
| 384 | 0.00 | 100.00 | 100.00 | 100.00 |

Table 1. Percentage of Snapshots with Energy Level Higher than Quantization Step

The interharmonic voltage-current correlation for all the locations is calculated as well, and shown in Table 2. The results obtained for the feeder show that its measurements may not reliable. For the loads, it can be seen that the correlation is generally high, except for that of Customer 2 at 264 Hz.

| IH freq (Hz) | Feeder | Customer 1 | Customer 2 | Customer 3 |
|---|---|---|---|---|
| 228 | 0.68 | 0.98 | 0.96 | 0.98 |
| 264 | 0.31 | 0.96 | 0.59 | 0.92 |
| 348 | 0.55 | 0.96 | 0.92 | 0.96 |
| 384 | 0.77 | 0.99 | 0.98 | 0.97 |

Table 2. $V$-$I$ Correlation Coefficient (%)

The other reliability criteria are also used but do not add much information to the conclusions to be drawn in Table 3, which summarizes the reliability at each frequency for each location.

| IH freq (Hz) | Feeder | Customer 1 | Customer 2 | Customer 3 |
|---|---|---|---|---|
| 228 | No | Yes | Yes | Yes |
| 264 | No | No | No | Yes |
| 348 | No | No | Yes | No |
| 384 | No | Yes | Yes | Yes |

Table 3. Reliability Summary

Table 4 shows the average of calculated active power at the feeder and at the loads (phase A). Note that the shaded cells are the ones that should not be trusted.

| IH Freq (Hz) | P (W) Feeder | P (W) Customer 1 | P (W) Customer 2 | P (W) Customer 3 |
|---|---|---|---|---|
| 228 | -0.52 | 0.09 | -3.53 | 0.72 |
| 264 | -1.14 | 0.07 | 0.13 | -0.75 |
| 348 | -1.05 | 0.08 | -1.98 | -0.14 |
| 384 | -6.34 | 0.58 | -23.5 | 3.64 |

Table 4. Active Power Results for the Feeder and Customers

## 6.2 The $V_{IH}$-$I_{IH}$ angle displacement

The power direction method relies on the information about the difference between the interharmonic voltage and current angles. If this difference is close to 90 or 270 degrees, the cosine of this difference will be very close to zero. For interharmonics of very low magnitude, the power may oscillate around zero, because the angle displacement usually exhibit lots of fluctuation due to measurement inaccuracies. Therefore, caution is needed when using the power direction method, since it is too sensitive to this angle.

In the present case study, such fluctuation happens for interharmonics 264 Hz and 348 Hz. Furthermore, the active power results shown in Fig. 13 and Fig. 14 reveal that the power level is very low. This was also shown in Table 1, which revealed that many snapshots contain data with very low energy level. For these frequencies, the conclusions drawn using the power direction method cannot be trusted. A final conclusion about these frequencies will be provided in next subsection by using the theory of interharmonic pairing.

## 6.3 The Interharmonic phase sequence characteristics

Using the phase sequence characteristics of interharmonics, it can be verified that interharmonics 228 Hz and 348 Hz of this case study are one pair, and interharmonics 264 Hz and 384 Hz are another pair. From (12), it can be estimated that the drives' frequencies are 48 Hz and 54 Hz, and that the number of pulses of the inverter is $p_2 = 6$. From this equation, it was also identified that 228 Hz and 264 Hz are negative sequence, whereas 348 Hz and 384 Hz are positive sequence, as explained in (Zhang et al, 2005). Therefore, all parameters in (12) can be estimated as

$$
\begin{aligned}
-228 &= 60 - 6 \times 48, \\
-364 &= 60 - 6 \times 54, \\
+348 &= 60 + 6 \times 48, \\
+384 &= 60 + 6 \times 54.
\end{aligned}
\tag{14}
$$

The same conclusion about the sequence is verified through analyzing the measurements: the symmetrical components of the interharmonic currents are calculated and one of them (positive-, negative- or zero-sequence) is observed to match the phase currents (the system is fairly balanced).

Since it is clear that the source of two interharmonic frequencies of a pair is the same, it is confirmed that Table 4 shows some inconsistencies: Customer 3 cannot be the source of interharmonic 264 Hz unless it is also the source of interharmonic 384 Hz. It was, however, determined that Customer 2 is the source of interharmonic 384 Hz. This inconsistency for Customer 2 undermines the credibility of the conclusions taken at this frequency. It is not possible that interharmonic 264 Hz comes from both Customer 3 and Customer 2. Finally the possibility that Customer 3 is the source of the interharmonic 264 Hz can be ruled out because this frequency is a pair of 384 Hz, which was generated from Customer 2.

## 7. Interharmonic source determination case study #2

In a second case, interharmonic problems were experienced in another oilfield area of Alberta, Canada. Measurements were taken at three customers, codenamed Customer 1, Customer 2, and Customer 3, which were operating big oil extraction ASD drives and were suspected interharmonic sources. The system diagram is shown in Fig. 16. The measurements at the metering points revealed that the interharmonic detected frequencies were present throughout the system.

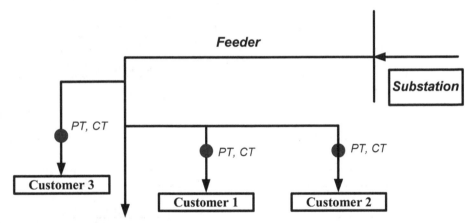

Fig. 16. Field measurement locations at system #2

Fig. 17 shows a sample contour plot of the spectrum calculated for the three Customers' currents in order to obtain the frequencies of the interharmonic components present in this system. Two main interharmonics are identified: 151 Hz and 271 Hz.

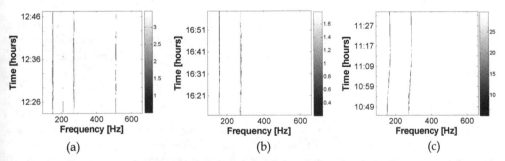

Figure 17. Contour plot of the interharmonic data recorded at the three Customers (phase A): (a) customer 1, (b) customer 2, (c) customer 3.

### 7.1 Criteria for determining the reliability of the data

Table 5 shows the percentage of reliable snapshots obtained by using the quantization error criterion. Only snapshots with an energy level higher than the quantization error could be used. All the data were reliable in this case due to the high magnitude of the interharmonic components.

| IH freq (Hz) | Customer 1 | Customer 2 | Customer 3 |
|:---:|:---:|:---:|:---:|
| 151 Hz | 100 | 100 | 100 |
| 271 Hz | 100 | 100 | 100 |

Table 5. Reliable Snapshots [%] According to the Quantization Error Criterion for System #2

Table 6 shows the correlation results for the locations, revealing that the interharmonic voltage and current had a high degree of correlation at the three customer loads.

| IH freq (Hz) | Customer 1 | Customer 2 | Customer 3 |
|:---|:---|:---|:---|
| 151 Hz | 0.93 | 0.88 | 0.98 |
| 271 Hz | 1.00 | 1.00 | 1.00 |

Table 6. Correlation Results for System #2

## 7.2 Results

The power direction results are shown in Table 7. In this case, the data was deemed reliable by all reliability criteria.

| Location of measurements | IH frequency [Hz] | Sign($P_{IH}$) |
|:---:|:---:|:---:|
| Customer 1 | 151 | + |
|  | 271 | + |
| Customer 2 | 151 | + |
|  | 271 | + |
| Customer 3 | 151 | - |
|  | 271 | - |

Table 7. Power Direction (at Interharmonic Frequencies) Results for System #2

The information for sign($P_{IH}$) reveals that the sign($P_{IH}$) of Customer 3 is negative, so that Customer 3 was the source. In this case, the angle displacement between the voltage and current was not observed to fluctuate at around $\pm\pi/2$ radians. Therefore, the power direction method can be used with full confidence.

## 8. Conclusions

This chapter investigated the reliability of the data used for the power quality disturbances assessment. The main applications were to estimate the network harmonic impedance and to determine the interharmonic source. A set of criteria to state about the data reliability was presented. They consisted in proposing thresholds for the following parameters:
- Frequency-domain coherence;
- Time-domain correlation;
- Quantization error;
- Standard deviation;

For the network impedance estimation application, it has been found that the energy level for $\Delta I(f)$ is a useful data filtering, but for $\Delta V(f)$ it has been found that it does not really make any difference. The Coherence index does not reveal much information about unreliable measurements but clearly identifies the principal frequency components of the transient. Analyses carried out on the quantization error level demonstrated that quantization noise is substantial for high frequencies and that the measurements taken are not significantly affected by quantization noises. The suggested thresholds for data rejection used were determined through extensive experience with handling data and provided more accurate and dependable results. These thresholds can be further adjusted as new data are analyzed and experience is built to improve the engineering judgment.

For the interharmonic source detection, the power direction method is very sensitive to the typical low energy level of the interharmonic currents. It was observed that this low energy level affects the displacement angle between voltage and current, which may prevent using the method to conclude about some frequencies. Interharmonic pairing theory was used to draw a final conclusion for the smaller-magnitude interharmonics in the case study. For the higher magnitude interharmonics, the power direction method could be used with the confidence provided by the reliability criteria.

## 9. Acknowledgments

The author gratefully acknowledges the financial support provided by Hydro One to partially cover the publication of this chapter and also their encouragement into improving this research work.

The author also expresses his gratitude towards Mr. Edwin Enrique Nino from ATCO electric for providing the several data sets of the network harmonic impedance collected in many Canadian utility sites, for processing the data and providing the presented results. Thanks are also extended to Mr. Hooman Mazin for providing part of the statistical analysis used in the network harmonic impedance data filtering application.

Finally, the author expresses his appreciation to Prof. Wilsun Xu for his supervision, technical contributions and high-level advices.

## 10. References

Anis, W. R. & Morcos, M. M. (2002). Artificial Intelligence and Advanced Mathematical Tools for Power Quality Applications: A Survey, *IEEE Trans. Power Delivery*, vol. 17, no. 2.

Axelberg, P. G. V., Bollen, M. J., Gu, I. Y. (2008). Trace of Flicker Sources by Using the Quantity of Flicker Power, *IEEE Trans. Power Delivery*, vol. 23, no. 1, pp. 465-471.

Ghartemani, M. K. & Iravani, M. R. (2005). Measurement of Harmonics/Interharmonics of Time-Varying Frequencies, *IEEE Trans. Power Delivery*, vol. 20, no. 1, pp. 23-31.

Harnett, D. L. (1982). Statistical Methods, Third Edition, Addison Wesley.

IEEE Std. 519-1992 (1993) *IEEE Recommended Practices and Requirements for Harmonic Control in Electrical Power Systems*

IEEE Std. 1159-1995. (1995). *IEEE Recommended Practice for Monitoring Electric Power Quality.*

IEEE Task Force on Harmonics Modeling and Simulation. (2007). Interharmonics: Theory and Measurement, *IEEE Trans. Power Delivery*, vol. 22, no. 4.

IEEE Interharmonic Task Force. (1997). *Cigre 36.05/CIRED 2 CC02 Voltage Quality Working Group, Interharmonics in Power Systems*.

Johnson, R. A. & Wichern, D. W. (2007). Applied Multivariate Statistical Analysis, Sixth edition, Pearson Prentice Hall.

Kim, T., Powers, E. J., Grady, W. M., Arapostathis, A. (2005). Real and Reactive Power Analysis for Interharmonics, *Proceedings of the IEEE Electric Ship Technologies Symposium*, pp. 244-247.

Kimbark, E. (1971). *Direct Current Transmission* (vol. 1), John Wiley & Sons

Li, C., Xu, W., Tayjasanant, T. (2001). Interharmonics: Basic Concepts and Techniques for their Detection and Measurement, *Electric Power Syst. Res.*, vol. 66, pp. 39-48.

Morched, A. S. and Kundur, P. (1987). Identification and Modeling of Load Characteristics at High Frequencies, *IEEE Trans. Power Systems*, vol. PWRS-2, no. 1, pp. 153-160.

Nassif, A. B, Yong, J., Mazin, H., Wang, X., Xu, W., (2010a). An Impedance-Based Approach for Identifying Interharmonic Sources. *IEEE Transactions on Power Delivery*, vol. 26, no. 1, pp. 333-340.

Nassif, A. B, Nino, E. E., Mazin, H. (2009). On the Reliability of Real Measurement Data for Assessing Power Quality Disturbances, *Proceedings of the IEEE PES General Meeting*.

Nassif, A. B., Yong, J., Xu, W. (2010b). Interharmonics: Signaling processing issues and applications, *Proceedings of the IEEE PES General Meeting*.

Oppenheim, A. V. and Schafer, R. W. (1999). *Discrete-Time Signal Processing*, 2nd Ed. (Prentice Hall: Upper Saddle River, NJ).

Oliveira, A., Oliveira, J. C., Resende J. W., Miskulin M. S. (1991). Practical approaches for AC system harmonic impedance measurements, *IEEE Trans. Power Delivery*, vol. 6, pp. 1721-1726.

Robert, A. & Deflandre, T. (1997). Guide for assessing the network harmonic impedances, Proceedings of the *Joint CIGRE/CIRED 97, 2-5 June 1997, Conference Publication No. 438*.

Sabin, D. D., Grebe, T. E., Brooks, D. L., Sundaram A. (1999), Rules-based Algorithm for Detecting Transient Overvoltages due to Capacitor Switching and Statistical Analysis of Capacitor Switching in Distribution Systems, *Proceedings of the IEEE Transmission and Distribution Conference*, vol. 2, pp. 630-635.

Xu, W., Ahmed, E. E., Zhang, X. and Liu, X. (2002). Measurement of Network Harmonic Impedances: Practical Implementation Issues and Their Solutions, *IEEE Trans. Power Delivery*, pp 210-216 vol. 17, no. 1.

Yacamini, R. (1996). Power System Harmonics: Part 4 Interharmonics, *IEE Power Eng. J.*, pp. 185-193.

Zhang, D., Xu W., Liu, Y. (2005). On the Phase Sequence Characteristics of Interharmonics, *IEEE Trans. Power Delivery*, vol. 20, no. 4, pp. 2563-2569.

# Part 2

# Converters

# 4

# Thermal Analysis of Power Semiconductor Converters

Adrian Plesca

*Gheorghe Asachi Technical University of Iasi*

*Romania*

## 1. Introduction

Power devices may fail catastrophically if the junction temperature becomes high enough to cause thermal runaway and melting. A much lower functional limit is set by temperature increases that result in changes in device characteristics, such as forward breakover voltage or the recovery time, and failure to meet device specifications.

Heat generation occurs primarily within the volume of the semiconductor pellet. This heat must be removed as efficiently as possible by some form of thermal exchange with the ambient, by the processes of conduction, convection or radiation.

Heat loss to the case and heat-sink is primarily by conduction. Heat loss by radiation accounts for only 1-2% of the total and can be ignored in most situations. Finally, loss from the heat-sink to the air is primarily by convection. When liquid cooling is used, the heat loss is by conduction to the liquid medium through the walls of the heat exchanger. Heat transfer by conduction is conveniently described by means of an electrical analogy, as it shows in Table 1.

| THERMAL | | | ELECTRICAL | | |
|---|---|---|---|---|---|
| Quantity | Symbol | Measure unit | Quantity | Symbol | Measure unit |
| Loss power | $P$ | W | Electric current | $I$ | A |
| Temperature variation | $\Delta\theta$ | $^0C$ | Voltage | $U$ | V |
| Thermal resistance | $R_{th}$ | $^0C/W$ | Electrical resistance | $R$ | $\Omega$ |
| Thermal capacity | $C_{th}$ | $J/^0C$ | Electrical capacity | $C$ | F |
| Heat | $Q$ | J | Electrical charge | $Q$ | As |
| Thermal conductivity | $\lambda$ | $W/m^0C$ | Electrical conductivity | $\sigma$ | $1/\Omega m$ |

Table 1. Thermal and electrical analogy

Taking into account the thermal phenomena complexity for power semiconductor devices it is very difficult to study the heating processes both in steady-state or transitory operating conditions, using the traditional analytical equations. The modeling concepts have their strength for different grades of complexity of the power circuit. It is important to achieve an efficient tradeoff between the necessary accuracy, required simulation speed and feasibility of parameter determination, (Kraus & Mattausch, 1998). Approaches to simulate these processes have already been made in earlier work. Numerical programs based on the method of finite differences are proposed in (Wenthen, 1970), or based on formulation of charge carrier transport equations, (Kuzmin et al., 1993). A physical model using the application of continuity equation for description of the carrier transport in the low doped layer of structures is proposed in (Schlogl et al., 1998). A simple calculation procedure for the time course of silicon equivalent temperature in power semiconductor components based on the previously calculated current loading is shown in (Sunde et al., 2006). In order to take into account the nonlinear thermal properties of materials a reduction method based on the Ritz vector and Kirchoff transformation is proposed in (Gatard et al., 2006).

The work described in (Chester & Shammas, 1993) outlines a model which combines the temperature dependent electrical characteristics of the device with its thermal response. The most papers are based on the thermal RC networks which use the PSpice software, (Maxim et al., 2000; Deskur & Pilacinski, 2005). In (Nelson et al., 2006) a fast Fourier analysis to obtain temperature profiles for power semiconductors is presented. Electro-thermal simulations using finite element method are reported in (Pandya & McDaniel, 2002) or combination with the conventional RC thermal network in order to obtain a compact model is described in (Shammas et al., 2002). Most of the previous work in this field of thermal analysis of power semiconductors is related only to the power device alone. But in the most practical applications, the power semiconductor device is a part of a power converter (rectifier or inverter). Hence, the thermal stresses for the power semiconductor device depend on the structure of the power converter. Therefore, it is important to study the thermal behaviour of the power semiconductor as a component part of the converter and not as an isolated piece. In the section 2, the thermal responses related to the junction temperatures of power devices have been computed. Parametric simulations for transient thermal conditions of some typical power rectifiers are presented in section 3. In the next section, the 3D thermal modelling and simulations of power device as main component of power converters are described.

## 2. Transient thermal operating conditions

The concept of thermal resistance can be extended to thermal impedance for time-varying situations. For a step of input power the transient thermal impedance, $Z_{thjCDC}(t)$, has the expression,

$$Z_{thjCDC}(t) = \frac{\Delta\theta_{jC}(t)}{P}$$
(1)

where:
$Z_{thjCDC}(t)$ means junction-case transient thermal impedance;
$\Delta\theta_{jC}(t)$ – difference of temperature between junction and case at a given time t;
$P$ – step of input power.
The transient thermal impedance can be approximated through a sum of exponential terms, like in expression bellow,

$$Z_{thjCDC}(t) = \sum_{j=1}^{k} r_j \left( 1 - e^{-\frac{t}{\tau_j}} \right) \tag{2}$$

where $\tau_j = r_j C_j$ means thermal time constant.

The response of a single element can be extended to a complex system, such as a power semiconductor, whose thermal equivalent circuit comprises a ladder network of the separate resistance and capacitance terms shown in Fig. 1.

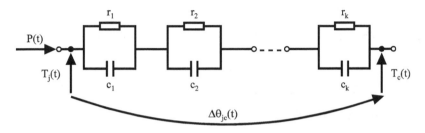

Fig. 1. Transient thermal equivalent circuit for power semiconductors

The transient response of such a network to a step of input power takes the form of a series of exponential terms. Transient thermal impedance data, derived on the basis of a step input of power, can be used to calculate the thermal response of power semiconductor devices for a variety of one-shot and repetitive pulse inputs. Further on, the thermal response for commonly encountered situations have been computed and are of great value to the circuit designer who must specify a power semiconductor device and its derating characteristics.

### 2.1 Rectangular pulse series input power

Figure 2 shows the rectangular pulse series and the equation (3) describes this kind of input power.

$$P(t) = \begin{cases} P_{FM} & if \quad nT \leq t \leq nT + \theta, \\ 0 & if \quad nT + \theta < t \leq (n+1)T \end{cases} \tag{3}$$

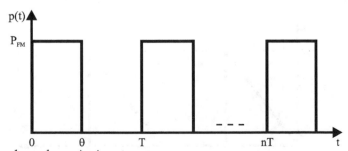

Fig. 2. Rectangular pulse series input power

The thermal response is given by the following equation,

$$
\Delta\theta_{jC(n+1)}(t)=\begin{cases}
P_{FM}\sum_{i=1}^{k}r_i\left[1-\dfrac{e^{\frac{t-nT}{T_i}}\left(1-e^{-\frac{(n+1)T}{T_i}}\right)-\left(1-e^{-\frac{nT}{T_i}}\right)e^{-\frac{T-\theta}{T_i}}}{1-e^{-\frac{T}{T_i}}}\right] & if\quad nT\le t\le nT+\theta, \\[2em]
P_{FM}\sum_{i=1}^{k}r_ie^{-\frac{t-nT-\theta}{T_i}}\dfrac{1-e^{-\frac{\theta}{T_i}}}{1-e^{-\frac{T}{T_i}}}\left(1-e^{-\frac{(n+1)T}{T_i}}\right) & if\quad nT+\theta<t\le(n+1)T
\end{cases}
\tag{4}
$$

For a very big number of rectangular pulses, actually $n\to\infty$, it gets the relation:

$$
\Delta\theta_{jC\infty}(t)=\begin{cases}
P_{FM}\sum_{i=1}^{k}r_i\left(1-e^{-\frac{t}{T_i}}\right)\dfrac{1-e^{-\frac{T-\theta}{T_i}}}{1-e^{-\frac{T}{T_i}}} & if\quad nT\le t\le nT+\theta, \\[2em]
P_{FM}\sum_{i=1}^{k}r_ie^{-\frac{t-\theta}{T_i}}\dfrac{1-e^{-\frac{\theta}{T_i}}}{1-e^{-\frac{T}{T_i}}} & if\quad nT+\theta<t\le(n+1)T
\end{cases}
\tag{5}
$$

Therefore, the junction temperature variation in steady-state conditions will be,

$$
\Delta\theta_{jC\infty}=\left(P_{FM}-\frac{\theta}{T}P_{FM}\right)\sum_{i=1}^{k}r_i-P_{FM}\sum_{i=1}^{k}r_ie^{-\frac{\theta}{T_i}}\frac{1-e^{-\frac{T-\vartheta}{T_i}}}{1-e^{-\frac{T}{T_i}}}=P_{FM}\sum_{i=1}^{k}r_i\left[1-\frac{\theta}{T}-e^{-\frac{\theta}{T_i}}\frac{1-e^{-\frac{T-\vartheta}{T_i}}}{1-e^{-\frac{T}{T_i}}}\right]=
$$

$$
=P_{FM}\sum_{i=1}^{k}r_i\left(\frac{1-e^{-\frac{\vartheta}{T_i}}}{1-e^{-\frac{T}{T_i}}}-\frac{\theta}{T}\right)
\tag{6}
$$

## 2.2 Increasing triangle pulse series input power

A series of increasing triangle pulses is shown in Fig. 3 and the equation which describes this series is given in (7).

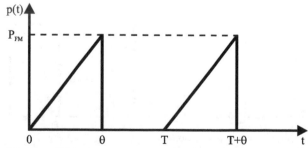

Fig. 3. Increasing triangle pulse series input power

$$P(t) = \begin{cases} t\dfrac{P_{FM}}{\theta} & if \quad nT \le t \le \theta + nT, \\ 0 & if \quad \theta + nT < t \le (n+1)T \end{cases} \tag{7}$$

In the case when $n \to \infty$, the thermal response will be,

$$\Delta\theta_{jC\infty}(t) = \begin{cases} \dfrac{P_{FM}}{\theta}\displaystyle\sum_{i=1}^{k} r_i\left\{ t - T_i\left[ 1 - \dfrac{1 - \left(1 - \dfrac{\theta}{T_i}\right)e^{-\frac{T-\theta}{T_i}}}{1 - e^{-\frac{T}{T_i}}}e^{-\frac{t}{T_i}} \right] \right\} & if \quad nT \le t \le \theta + nT, \\[4ex] \dfrac{P_{FM}}{\theta}\displaystyle\sum_{i=1}^{k} r_i T_i \dfrac{\dfrac{\theta}{T_i} + e^{-\frac{\theta}{T_i}} - 1}{1 - e^{-\frac{T}{T_i}}}e^{-\frac{t-\theta}{T_i}} & if \quad \theta + nT < t \le (n+1)T \end{cases} \tag{8}$$

## 2.3 Decreasing triangle pulse series input power

Figure 4 shows a decreasing triangle pulse series with its equation (9).

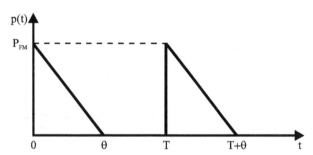

Fig. 4. Decreasing triangle pulse series input power

$$P(t) = \begin{cases} 1 - t\dfrac{P_{FM}}{\theta} & if \quad nT \le t \le \theta + nT, \\ 0 & if \quad \theta + nT < t \le (n+1)T \end{cases} \tag{9}$$

At limit, when $n \to \infty$, the thermal response will be:

$$\Delta\theta_{jC\infty}(t) = \begin{cases} \dfrac{P_{FM}}{\theta}\displaystyle\sum_{i=1}^{k} r_i\left\{ (\theta - t) - T_i\left[ -1 + \dfrac{1 + \dfrac{\theta}{T_i} - e^{-\frac{T-\theta}{T_i}}}{1 - e^{-\frac{T}{T_i}}}e^{-\frac{t}{T_i}} \right] \right\} & if \quad nT \le t \le \theta + nT, \\[4ex] \dfrac{P_{FM}}{\theta}\displaystyle\sum_{i=1}^{k} r_i T_i \dfrac{1 - \left(1 + \dfrac{\theta}{T_i}\right)e^{-\frac{\theta}{T_i}}}{1 - e^{-\frac{T}{T_i}}} & if \quad \theta + nT < t \le (n+1)T \end{cases} \tag{10}$$

## 2.4 Triangle pulse series input power

A series of triangle input power is shown in Fig. 5. The equation which describes this kind of series is given in (11).

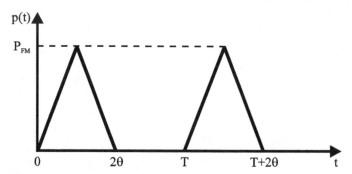

Fig. 5. Triangle pulse series input power.

$$P(t) = \begin{cases} t\dfrac{P_{FM}}{\theta} & if \quad nT \le t \le \theta + nT, \\[2ex] \left(2 - \dfrac{t}{\theta}\right)P_{FM} & if \quad \theta + nT < t \le 2\theta + nT, \\[2ex] 0 & if \quad 2\theta + nT < t \le (n+1)T \end{cases} \tag{11}$$

For junction temperature computation when $n \to \infty$, the following relation will be used:

$$\Delta\theta_{jC\infty}(t) = \begin{cases} \dfrac{P_{FM}}{\theta}\displaystyle\sum_{i=1}^{k} r_i \left[ T_i \dfrac{1 - 2e^{\frac{T-\theta}{T_i}} + e^{\frac{T-2\theta}{T_i}}}{1 - e^{\frac{T}{T_i}}} e^{-\frac{t}{T_i}} + (t - T_i) \right] & if \quad nT \le t \le \theta + nT, \\[5ex] \dfrac{P_{FM}}{\theta}\displaystyle\sum_{i=1}^{k} r_i \left\{ -T_i \dfrac{2 - e^{\frac{\theta}{T_i}} - e^{\frac{T-\theta}{T_i}}}{1 - e^{\frac{T}{T_i}}} e^{-\frac{t-\theta}{T_i}} + \left[(2\theta - t) + T_i\right] \right\} & if \quad \theta + nT < t \le 2\theta + nT, \\[5ex] \dfrac{P_{FM}}{\theta}\displaystyle\sum_{i=1}^{k} r_i T_i \dfrac{\left(1 - e^{-\frac{\theta}{T_i}}\right)^2}{1 - e^{\frac{T}{T_i}}} e^{-\frac{t-2\theta}{T_i}} & if \quad 2\theta + nT < t \le (n+1)T \end{cases} \tag{12}$$

## 2.5 Trapezoidal pulse series input power

Figure 6 shows a trapezoidal pulse series with the equation from (13).

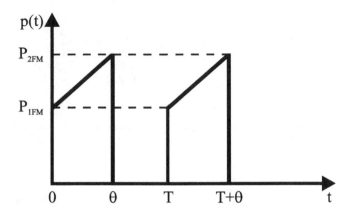

Fig. 6. Trapezoidal pulse series input power

$$P(t) = \begin{cases} P_{1FM} + (P_{2FM} - P_{1FM})\dfrac{t}{\theta} & if \quad nT \le t \le \theta + nT, \\ 0 & if \quad \theta + nT < t \le (n+1)T \end{cases} \tag{13}$$

At limit, $n \to \infty$, the thermal response is given by,

$$\Delta\theta_{jC\infty}(t) = \begin{cases} \dfrac{1}{\theta}\displaystyle\sum_{i=1}^{k} r_i \left[ T_i \dfrac{F_{P_1 P_2}}{1 - e^{-\frac{T}{T_i}}} e^{-\frac{t}{T_i}} + P_{1FM} + \dfrac{P_{2FM} - P_{1FM}}{\theta}(t - T_i) \right] & if \quad nT \le t \le \theta + nT, \\[4mm] \dfrac{1}{\theta}\displaystyle\sum_{i=1}^{k} r_i T_i \dfrac{G_{P_1 P_2}}{1 - e^{-\frac{T}{T_i}}} e^{-\frac{t-\theta}{T_i}} & if \quad \theta + nT < t \le (n+1)T \end{cases} \tag{14}$$

where:

$$F_{P_1 P_2} = P_{2FM} - P_{1FM}\left(1 + \dfrac{\theta}{T_i}\right) + \left[ P_{1FM} - P_{2FM}\left(1 - \dfrac{\theta}{T_i}\right) \right] e^{-\frac{T-\theta}{T_i}},$$

$$G_{P_1 P_2} = P_{1FM} - P_{2FM}\left(1 - \dfrac{\theta}{T_i}\right) + \left[ P_{2FM} - P_{1FM}\left(1 + \dfrac{\theta}{T_i}\right) \right] e^{-\frac{\theta}{T_i}} \tag{15}$$

## 2.6 Partial sinusoidal pulse series input power

A partial sinusoidal pulse series waveform is shown in Fig. 7. The equation which describes this kind of waveform is given by (16).

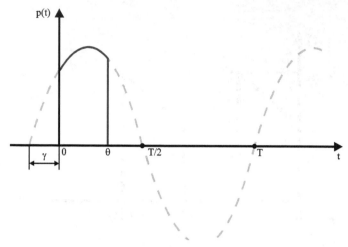

Fig. 7. Partial sinusoidal pulse series input power

$$P(t) = \begin{cases} P_{FM}\sin(\omega t + \gamma) & if \quad nT \le t \le \theta + nT, \\ 0 & if \quad \theta + nT < t \le (n+1)T \end{cases} \tag{16}$$

In order to establish the junction temperature when $n \to \infty$, it will use the relation,

$$\Delta\theta_{jC\infty}(t) = \begin{cases} P_{FM}\left\{ Z\sin(\omega t + \gamma - \delta) - \sum_{i=1}^{k} r_i \left[ \sin(\gamma - \varphi_i) - \sin(\gamma - \varphi_i + \omega\theta)e^{-\frac{T-\theta}{T_i}} \right] \dfrac{e^{-\frac{t}{T_i}}}{\left(1 - e^{-\frac{T}{T_i}}\right)\sqrt{1 + (\omega T_i)^2}} \right\} \\ \qquad if \quad nT \le t \le \theta + nT, \\ P_{FM}\sum_{i=1}^{k} r_i \left[ \sin(\omega\theta + \gamma - \varphi_i) - \sin(\gamma - \varphi_i)e^{-\frac{\theta}{T_i}} \right] \dfrac{e^{-\frac{t-\theta}{T_i}}}{\left(1 - e^{-\frac{T}{T_i}}\right)\sqrt{1 + (\omega T_i)^2}} \\ \qquad if \quad \theta + nT < t \le (n+1)T \end{cases} \tag{17}$$

where:

$$ctg\phi_i = \frac{1}{\omega T_i}; \quad Z^2 = \sum_{i=1}^{k}\left(r_i\cos^2\varphi_i\right)^2 + \sum_{i=1}^{k}\left(\frac{r_i}{2}\sin 2\varphi_i\right)^2; \quad tg\delta = \frac{\displaystyle\sum_{i=1}^{k}\frac{r_i}{2}\sin 2\varphi_i}{\displaystyle\sum_{i=1}^{k}r_i\cos^2\varphi_i} \tag{18}$$

Extremely short overloads of the type that occur under surge or fault conditions, are limited to a few cycles in duration. Here the junction temperature exceeds its maximum rating and all operational parameters are severely affected. However the low transient thermal impedance offered by the device in this region of operation, is often sufficient to handle the power that is dissipated.

A transient thermal calculation even using the relation (2), is very complex and difficult to do. Hence, a more exactly and efficiently thermal calculation of power semiconductors at different types of input power specific to power converters, can be done with the help of PSpice software and/or 3D finite element analysis.

## 3. Thermal simulations of power semiconductors from rectifiers

Further on, it presents the waveforms of input powers and junction temperatures of power semiconductors, diodes and thyristors, from different types of single-phase bridge rectifiers. Also, temperature waveforms in the case of steady state thermal conditions, are shown. Using PSpice software, a parametric simulation which highlights the influence of some parameter values upon temperature waveforms has been done.

On ordinate axis, the measurement unit in the case of input power waveforms, is the watt, and in the case of temperatures, the measurement unit is the $^0$C, unlike the volt one that appears on graphics. This apparent unconcordance between measurement units is because thermal phenomena had been simulated using electrical circuit analogy. The notations on the graphics $P_1$, $P_2$ and $P_3$ mean input powers and $T_1$, $T_2$ and $T_3$ temperatures, respectively.

### 3.1 Single-phase uncontrolled bridge rectifier

The waveforms of the input powers and junction temperatures of power diodes from the structure of a single-phase uncontrolled bridge rectifier are shown in the below diagrams.

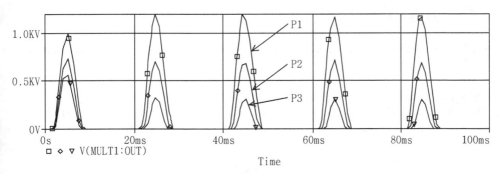

Fig. 8. Input power waveforms at load resistance variation with 10, 20, 50$\Omega$

From the above graphics, Fig. 8, the input power variation $P_1$, $P_2$ and $P_3$ with the load resistance values can be noticed. The increase of load values leads to small input power values, and finally, to the decrease of junction temperature magnitudes, $T_1$, $T_2$ and $T_3$, Fig. 9, and also to the decrease of temperature variations. In the case of quasi-steady state thermal conditions, Fig. 10, there are a clearly difference between temperatures waveforms variation. Also, the time variations of temperature values are insignificantly. The maximum value of $T_1$ temperature, Fig. 10, outruns the maximum admissible value for power semiconductor junction, about 125$^0$C. Therefore, it requires an adequate protection for the power diode or increasing of load resistance.

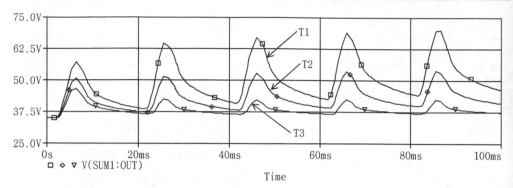

Fig. 9. Temperature waveforms of thermal transient conditions at load variation with 10, 20, 50Ω

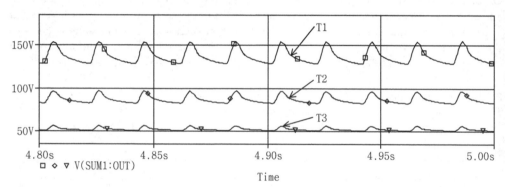

Fig. 10. Temperature waveforms of quasi-steady state thermal conditions at load variation with 10, 20, 50Ω

### 3.2 Single-phase semicontrolled bridge rectifier

In the case of a single-phase semicontrolled bridge rectifier made with power diodes and thyristors, the time variations of input powers and temperatures are presented below.

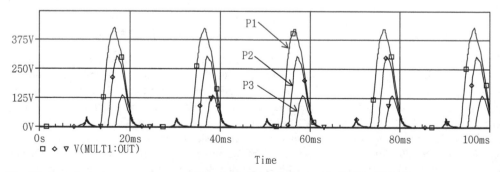

Fig. 11. Input power waveforms at firing angle variation with 60, 90, 120° el.

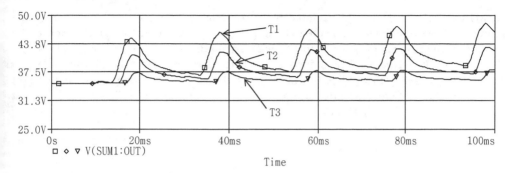

Fig. 12. Temperature waveforms of thermal transient conditions at firing angle variation with 60, 90, 120⁰ el.

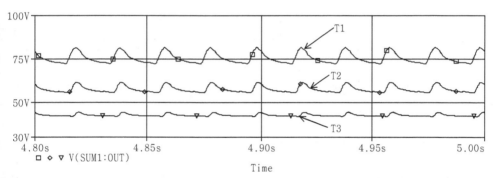

Fig. 13. Temperature waveforms of quasi-steady state thermal conditions at firing angle variation with 60, 90, 120⁰ el.

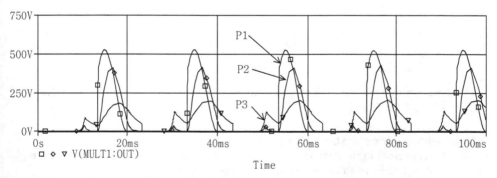

Fig. 14. Input power waveforms at load inductance variation with 0.1, 10, 50mH

It has been done a parametric simulation both at firing angle variation of thyristors from semicontrolled bridge rectifier, Fig. 11...13, and at load inductance variation, Fig. 14...16.
As in previous situation, the case with uncontrolled bridge rectifier, the variation of input power values depend on load inductance, Fig. 14. The increase of inductance value, from

0.1mH to 50mH, leads not only to input power decreasing, $P_3 < P_2 < P_1$, but also its shape changing. The same thing can be observed at firing angle variation, Fig. 11...13. Hence, the increase of the firing angle from 60 to 120⁰ el., leads to decrease of input power values $P_3 < P_2 < P_1$. Also, the increase of load inductance leads to decrease of temperature values, $T_3 < T_2 < T_1$, as shown in Fig. 15 and Fig. 16. The steady state thermal conditions allow to highlight the temperature differences in the case of firing angle variation, $T_3 < T_2 < T_1$, Fig. 13, and load variation, $T_3 < T_2 < T_1$, Fig. 16.

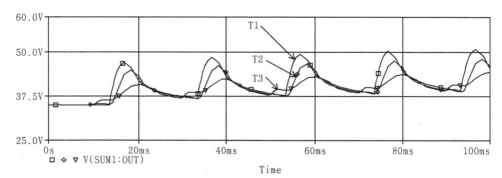

Fig. 15. Temperature waveforms of thermal transient conditions at load inductance variation with 0.1, 10, 50mH

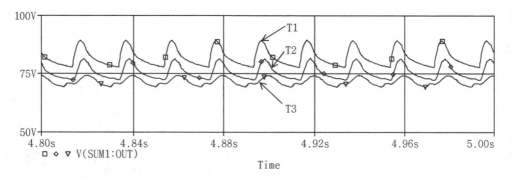

Fig. 16. Temperature waveforms of quasi-steady state thermal conditions at load inductance variation with 0.1, 10, 50mH

### 3.3 Single-phase controlled bridge rectifier

Next diagrams present input power variation and temperature values in the case of a single-phase controlled bridge rectifier made with power thyristors.

As in the case of single-phase semicontrolled bridge rectifier, a parametric simulation for firing angle variation has been done. It can be noticed that increasing of firing angle leads to input power and temperature decrease, Fig. 17 and Fig. 18. The quasi-steady state thermal conditions highlight the differences between temperature values and their variations, Fig. 19. In order to validate the thermal simulations some experimental tests have been done. It was recorded the temperature rise on the case of the thyristors used for semi-controlled power rectifier. The temperatures have been measured using proper iron-constantan thermocouples fixed on the case of power semiconductor devices. The measurements have been done both for the firing angle values of 60, 90 and 120⁰ el., and load inductance values of 0.1, 10 and 50mH. The results are shown in Fig. 20 and Fig. 21.

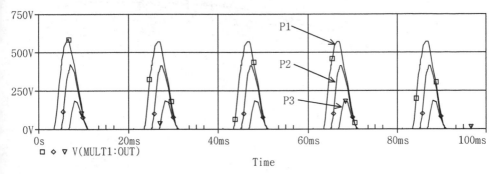

Fig. 17. Input power waveforms at firing angle variation with 60, 90, 120⁰ el.

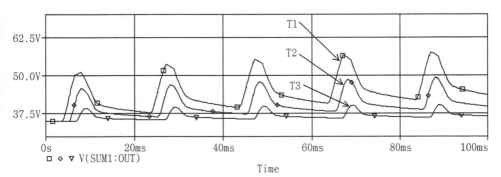

Fig. 18. Temperature waveforms of thermal transient conditions at firing angle variation with 60, 90, 120⁰ el.

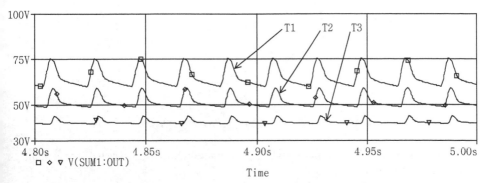

Fig. 19. Temperature waveforms of quasi-steady state thermal conditions at firing angle variation with 60, 90, 120⁰ el.

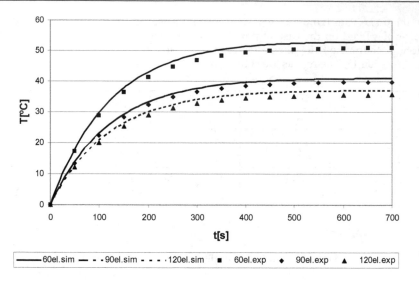

Fig. 20. Comparison between simulation and experimental temperature rise of the case at firing angle variation

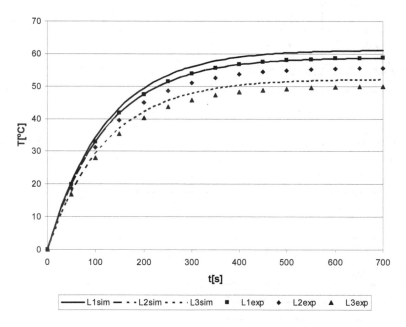

Fig. 21. Comparison between simulation and experimental temperature rise of the case at inductance load variation

In both cases, at firing angle and load inductance variation, it ca be noticed closer values between simulation results and measurements. Of course, there are different temperature values resulted from experimental tests (60el.exp, 90el.exp and 120el.exp from Fig. 20 and L1exp, L2exp, L3exp as shown in Fig. 21) with respect to simulations (60el.sim, 90el.sim and 120el.sim from Fig. 20 and L1sim, L2sim, L3sim as in Fig. 21), because of measurement errors, thermal model simplifications and mounting test conditions. Anyway, the maximum difference between experimental and simulation results is less than 3°C.

### 3.4 D thermal modelling and simulations of power semiconductors

During former work, (Chung, 1999; Allard et al., 2005), because of limited computer capabilities, the authors had to concentrate on partial problems or on parts of power semiconductors geometry. The progress in computer technology enables the modelling and simulation of more and more complex structures in less time. It has therefore been the aim of this work to develop a 3D model of a power thyristor as main component part from power semiconductor converters.

The starting point is the power balance equation for each volume element dV, in the integral formulation:

$$\iiint \frac{j^2}{\sigma} dV = \iiint \rho c \frac{\partial T}{\partial t} dV - \iiint div(\lambda \cdot gradT) dV \tag{19}$$

where:
T means the temperature of element [°C];
j – current density [A/m²];
σ – electrical conductivity [1/Ωm];
ρ – material density [kg/m³];
c – specific heat [J/kg°C];
λ – thermal conductivity [W/m°C].

The left term of before equation (it exists only in the device conductor elements), denotes the heating power from the current flow. It is in balance with the heat stored by temporal change of temperature, and the power removed from the element by thermal conduction. For the steady state temperature calculation, the heat storage term is zero, and the equation (19) becomes,

$$\iiint \frac{j^2}{\sigma} dV = -\iiint div(\lambda \cdot gradT) dV \tag{20}$$

Taking one's stand on the above thermal equations, first of all a 3D model for a power thyristor has been developed using a specific software, the Pro-ENGINEER, an integrated thermal design tool for all type of accurate thermal analysis on devices.

It has been considered an application which includes a bidirectional bridge equiped with power thyristors type AT505, with the average direct current of 430A and an internal resistance of 0.68mΩ. The current which flows through the converter branches is about 315A. This value allows computing the power loss for each tyristor, which results in 67.47W. The material properties of every component part of the thyristor are described in the Table 2 and the 3D thermal models of the thyristor with its main component parts and together with its heatsinks for both sides cooling are shown in Fig. 22, respectively, Fig. 23.

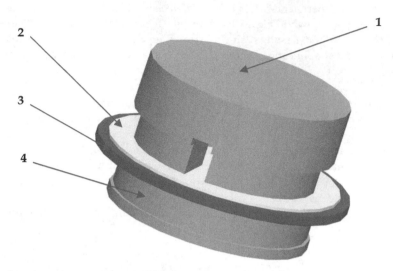

Fig. 22. Thermal model of the thyristor (1 – cathode copper pole; 2 – silicon chip; 3 – molybdenum disc; 4 – anode copper)

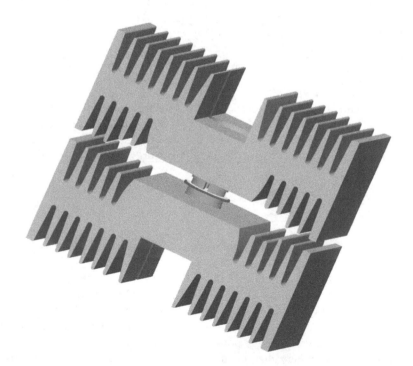

Fig. 23. Thermal model of the assembly thyristor - heatsinks

The thermal model of the power semiconductor has been obtained by including all the piece part that is directly involved in the thermal exchange phenomenon, which is: anode copper pole, molybdenum disc, silicon chip, cathode copper pole, Fig. 22. The device ceramic enclosure has not been included in the model since the total heat flowing trough it is by far less important than the heat flowing through the copper poles. All the mechanical details which are not important for the heat transfer within the thyristor and from the thyristor to the external environment (e.g. the centering hole on the poles) have been suppressed.

| Parameter | Material | | | |
| --- | --- | --- | --- | --- |
| | Copper | Silicon | Molybdenum | Aluminium |
| $\rho$ (kg/m$^3$) | 8900 | 2330 | 10220 | 2700 |
| c (J/kg°C) | 387 | 702 | 255 | 900 |
| $\lambda$ (W/m°C) | 385 | 124 | 138 | 200 |

Table 2. Material Data and Coefficients at 20°C

The heat load has been applied on the active surface of the silicon of power semiconductor. It is a uniform spatial distribution on this surface. The ambient temperature was about 25°C. From experimental tests it was computed the convection coefficient value, $k_t$ = 14.24W/m²°C for this type of heatsinks for thyristor cooling. Hence, it was considered the convection condition like boundary condition for the outer boundaries such as heatsinks. The convection coefficient has been applied on surfaces of heatsinks with a uniform spatial variation and a bulk temperature of 25°C. The mesh of this 3D power semiconductor thermal model has been done using tetrahedron solids element types with the following allowable angle limits (degrees): maximum edge: 175; minimum edge: 5; maximum face: 175; minimum face: 5. The maximum aspect ratio was 30 and the maximum edge turn (degrees): 95. Also, the geometry tolerance had the following values: minimum edge length: 0.0001; minimum surface dimension: 0.0001; minimum cusp angle: 0.86; merge tolerance: 0.0001. The single pass adaptive convergence method to solve the thermal steady-state simulation has been used.

Then, it has been made some steady-state thermal simulations for the power semiconductor. For all thermal simulations a 3D finite elements Pro-MECHANICA software has been used. The temperature distribution of the tyristor which uses double cooling, both on anode and cathode, is shown in the pictures below, Fig. 24 and Fig. 25. The maximum temperature for the power semiconductor is on the silicon area and is about 70.49°C and the minimum of 47.97°C is on the heatsink surfaces.

Further on, the thermal transient simulations have been done in order to compute the transient thermal impedance for power thyristor. The result is shown in Fig. 26.

From thermal transient simulations we obtain the maximum temperature time variation and the minimum temperature time variation. From the difference between maximum temperature time variation and ambient temperature divided to total thermal load it gets the thermal transient impedance. Dividing the thermal transient impedance to the thermal resistance, the normalised thermal transient impedance can be obtained. This is a thermal quantity which reflects the power semiconductor thermal behaviour during transient conditions.

To understand and to optimize the operating mechanisms of power semiconductor converters, the thermal behaviour of the power device itself and their application is of major interest. Having the opportunity to simulate the thermal processes at the power

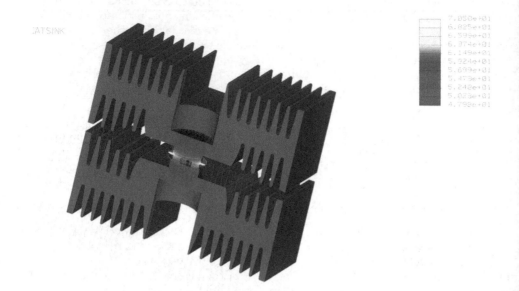

Fig. 24. Temperature distribution through the thyristor mounted between heatsinks at 50% cross section, yz plane

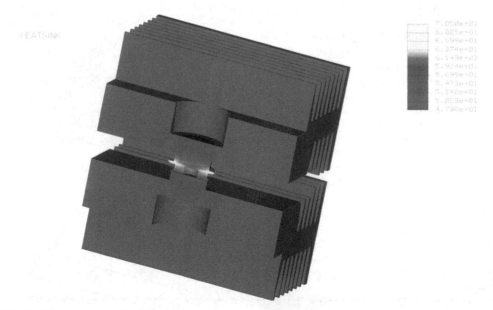

Fig. 25. Temperature distribution through the thyristor mounted between heatsinks at 50% cross section, xz plane

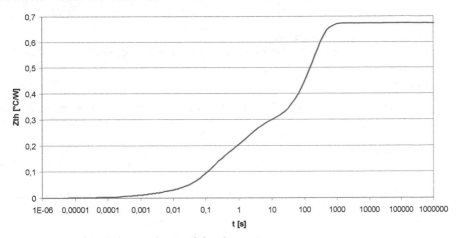

Fig. 26. Transient thermal impedance of the thyristor

semiconductor junction dependent on the power device design enables new features for the optimization of power semiconductor converters. This has a great impact to the development and test costs of new power converters.

## 4. Conclusion

From all previous thermal modelling, simulation and experimental tests, the following conclusions about transient thermal evolution of power semiconductor devices can be outlined:

- the shape of input power and temperatures evolution depend on load type, its value and firing angle in the case of power semicontrolled rectifiers;
- increasing of load inductance value leads to decrease of input power and temperature values;
- in the case of steady state thermal conditions, the temperature variation is not so important at big values of load inductance and firing angle;
- at big values of firing angle it can be noticed a decrease of input power values and temperatures;
- there is a good correlation between simulation results and experimental tests;
- because of very complex thermal phenomenon the analysis of power semiconductor device thermal field can be done using a specific 3D finite element method software; therefore, the temperature values anywhere inside or on the power semiconductor assembly can be computed both for steady-state or transient conditions;
- using the 3D simulation software there is the possibility to improve the power semiconductor converters design and also to get new solutions for a better thermal behaviour of power semiconductor devices.

Extending the model with thermal models for the specific applications enables the user of power semiconductors to choose the right ratings and to evaluate critical load cycles and to identify potential overload capacities for a dynamic grid loading. It was shown that the described thermal network simulation has a high potential for a variety of different applications:

- development support;

- identifying user risks;
- evaluating the right rated current;
- evaluating overload capacity without destructive failure of the power semiconductor.

## 5. References

Allard, B., Garrab, H. & Morel, H. (2005). Electro-thermal simulation including a temperature distribution inside power semiconductor devices, *International Journal of Electronics*, vol.92, pp. 189-213, ISSN 0020-7217

Chester, J. & Shammas, N. (1993). Thermal and electrical modelling of high power semiconductor devices, *IEE Colloquium on Thermal Management in Power Electronics Systems*, pp. 3/1 - 3/7, London, UK

Chung, Y. (1999). Transient thermal simulation of power devices with Cu layer, *Proc. 11th International Symposium on Power Semiconductor Devices and ICs. ISPSD'99*, pp. 257-260, ISBN 0-7803-5290-4

Deskur, J. & Pilacinski, J. (2005). Modelling of the power electronic converters using functional models of power semiconductor devices in Pspice, *European Conference on Power Electronics and Applications*, ISBN 90-75815-09-3

Gatard, E., Sommet, R. & Quere, R. (2006). Nonlinear thermal reduced model for power semiconductor devices, *Proc. 10th Intersociety Conference on Thermal and Thermomechanical Phenomena in Electronics Systems*, ISBN 0-7803-9524-7

Kraus, R. & Mattausch, H. (1998). Status and trends of power semiconductor device models for circuit simulation. *IEEE Transactions on Power Electronics*, vol.13, pp. 452 – 465, ISSN 0885-8993

Kuzmin, V., Mnatsakanov, T., Rostovtsev, I. & Yurkov, S. (1993). Problems related to power semiconductor device modelling, *Fifth European Conference on Power Electronics and Applications*, pp. 113 – 117, ISBN 0-8529-6587-7

Maxim, A., Andreu, D., & Boucher, J. (2000). A unified high accuracy SPICE library for the power semiconductor devices built with the analog behavioral macromodeling technique, *Proc. 12th Int. Symp. on Power Semiconductor Devices and Ics*, pp. 189 – 192, ISBN 0-7803-6269-1

Nelson, J., Venkataramanan, G. & El-Refaie, A. (2006). Fast thermal profiling of power semiconductor devices using Fourier techniques, *IEEE Transactions on Industrial Electronics*, vol.53, pp. 521 – 529, ISSN 0278-0046

Pandya, K. & McDaniel, W. (2002). A simplified method of generating thermal models for power MOSFETs, *Proc. Eighteenth Annual IEEE Semiconductor Thermal Measurement and Management Symposium*, ISBN 0-7803-7327-8

Schlogl, A., Mnatsakanov, T. & Schroder, D. (1998). Temperature dependent behaviour of silicon power semiconductors-a new physical model validated by device-internal probing between 400 K and 100 K, *Proc. of the 10th Int. Symp. on Power Semiconductor Devices and ICs ISPSD*, pp. 383 – 386, ISBN 0-7803-5100-2

Shammas, N., Rodriguez, M. & Masana, F. (2002). A simple evaluation method of the transient thermal response of semiconductor packages, *Microelectronics Reliability*, vol.42, pp. 109-117, ISSN 0026-2714

Sunde, V., Jakopovic, Z. & Cobanov, N. (2006). Simple Hybrid Electrothermal Simulation Procedure, *12th International Power Electronics and Motion Control Conference*, pp. 617 – 620, ISBN 1-4244-0121-6

Wenthen, F. (1970). Computer-aided thermal analysis of power semiconductor devices. *IEEE Transactions on Electron Devices*, vol.17, pp. 765 – 770, ISSN 0018-9383

# 5

# Study of LCC Resonant Transistor DC / DC Converter with Capacitive Output Filter

Nikolay Bankov, Aleksandar Vuchev and Georgi Terziyski
*University of Food Technologies – Plovdiv*
*Bulgaria*

## 1. Introduction

The transistor LCC resonant DC/DC converters of electrical energy, working at frequencies higher than the resonant one, have found application in building powerful energy supplying equipment for various electrical technologies (Cheron et al., 1985; Malesani et al., 1995; Jyothi & Jaison, 2009). To a great extent, this is due to their remarkable power and mass-dimension parameters, as well as, to their high operating reliability. Besides, in a very wide-working field, the LCC resonant converters behave like current sources with big internal impedance. These converters are entirely fit for work in the whole range from no-load to short circuit while retaining the conditions for soft commutation of the controllable switches.

There is a multitude of publications, dedicated to the theoretical investigation of the LCC resonant converters working at a frequency higher than their resonant one (Malesani et al., 1995; Ivensky et al. 1999). In their studies most often the first harmonic analysis is used, which is practically precise enough only in the field of high loads of the converter. With the decrease in the load the mistakes related to using the method of the first harmonic could obtain fairly considerable values.

During the analysis, the influence of the auxiliary (snubber) capacitors on the controllable switches is usually neglected, and in case of availability of a matching transformer, only its transformation ratio is taken into account. Thus, a very precise description of the converter operation in a wide range of load changes is achieved. However, when the load resistance has a considerable value, the models created following the method mentioned above are not correct. They cannot be used to explain what the permissible limitations of load change depend on in case of retaining the conditions for soft commutation at zero voltage of the controllable switches – zero voltage switching (ZVS).

The aim of the present work is the study of a transistor LCC resonant DC/DC converter of electrical energy, working at frequencies higher than the resonant one. The possible operation modes of the converter with accounting the influence of the damping capacitors and the parameters of the matching transformer are of interest as well. Building the output characteristics based on the results from a state plane analysis and suggesting a methodology for designing, the converter is to be done. Drawing the boundary curves between the different operating modes of the converter in the plane of the output characteristics, as well as outlining the area of natural commutation of the controllable switches are also among the aims of this work. Last but not least, the work aims at designing and experimental investigating a laboratory prototype of the LCC resonant converter under consideration.

## 2. Modes of operation of the converter

The circuit diagram of the LCC transistor resonant DC/DC converter under investigation is shown in figure 1. It consists of an inverter (controllable switches constructed on base of the transistors $Q_1 \div Q_4$ with freewheeling diodes $D_1 \div D_4$), a resonant circuit ($L$, $C$), a matching transformer $Tr$, an uncontrollable rectifier ($D_5 \div D_8$), capacitive input and output filters ($C_{F1}$ и $C_{F2}$) and a load resistor ($R_0$). The snubber capacitors ($C_1 \div C_4$) are connected with the transistors in parallel.

The output power of the converter is controlled by changing the operating frequency, which is higher than the resonant frequency of the resonant circuit.

It is assumed that all the elements in the converter circuit (except for the matching transformer) are ideal, and the pulsations of the input and output voltages can be neglected.

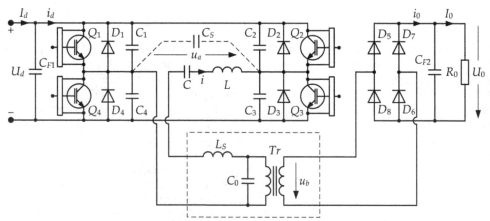

Fig. 1. Circuit diagram of the LCC transistor DC/DC converter

All snubber capacitors $C_1 \div C_4$ are equivalent in practice to just a single capacitor $C_S$ (dotted line in fig.1), connected in parallel to the output of the inverter. The capacity of the capacitor $C_S$ is equal to the capacity of each of the snubber capacitors $C_1 \div C_4$.

The matching transformer $Tr$ is shown in fig.1 together with its simplified equivalent circuit under the condition that the magnetizing current of the transformer is negligible with respect to the current in the resonant circuit. Then this transformer comprises both the full leakage inductance $L_S$ and the natural capacity of the windings $C_0$, reduced to the primary winding, as well as an ideal transformer with its transformation ratio equal to $k$.

The leakage inductance $L_S$ is connected in series with the inductance of the resonant circuit $L$ and can be regarded as part of it. The natural capacity $C_0$ takes into account the capacity between the windings and the different layers in each winding of the matching transformer. $C_0$ can has an essential value, especially with stepping up transformers (Liu et al., 2009).

Together with the capacity $C_0$ the resonant circuit becomes a circuit of the third order ($L$, $C$ and $C_0$), while the converter could be regarded as LCC resonant DC/DC converter with a capacitive output filter.

The parasitic parameters of the matching transformer – leakage inductance and natural capacity of the windings – should be taken into account only at high voltages and high operating frequencies of the converter. At voltages lower than 1000 V and frequencies lower than 100 kHz they can be neglected, and the capacitor $C_0$ should be placed additionally (Liu et al., 2009).

Because of the availability of the capacitor $C_S$, the commutations in the output voltage of the inverter ($u_a$) are not instantaneous. They start with switching off the transistors $Q_1/Q_3$ or $Q_2/Q_4$ and end up when the equivalent snubber capacitor is recharged from $+U_d$ to $-U_d$ or backwards and the freewheeling diodes $D_2/D_4$ or $D_1/D_3$ start conducting. In practice the capacitors $C_2$ and $C_4$ discharge from $+U_d$ to 0, while $C_1$ and $C_3$ recharge from 0 to $+U_d$ or backwards. During these commutations, any of the transistors and freewheeling diodes of the inverter does not conduct and the current flowed through the resonant circuit is closed through the capacitor $C_S$.

Because of the availability of the capacitor $C_0$, the commutations in the input voltage of the rectifier ($u_b$) are not instantaneous either. They start when the diode pairs ($D_5/D_7$ or $D_6/D_8$) stop conducting at the moments of setting the current to zero through the resonant circuit and end up with the other diode pair ($D_6/D_8$ или $D_5/D_7$) start conducting, when the capacitor $C_0$ recharges from $+kU_0$ to $-kU_0$ or backwards. During these commutations, any of the diodes of the rectifier does not conduct and the current flowed through the resonant circuit is closed through the capacitor $C_0$.

The condition for natural switching on of the controllable switches at zero voltage (ZVS) is fulfilled if the equivalent snubber capacitor $C_S$ always manages to recharge from $+U_d$ to $-U_d$ or backwards. At modes, close to no-load, the recharging of $C_S$ is possible due to the availability of the capacitor $C_0$. It ensures the flow of current through the resonant circuit, even when the diodes of the rectifier do not conduct.

When the load and the operating frequency are deeply changed, three different operation modes of the converter can be observed.

It is characteristic for the first mode that the commutations in the rectifier occur entirely in the intervals for conducting of the transistors in the inverter. This mode *is the main operation mode* of the converter. It is observed at comparatively small values of the load resistor $R_0$.

At the second mode the commutation in the rectifier ends during the commutation in the inverter, i.e., the rectifier diodes start conducting when both the transistors and the freewheeling diodes of the inverter are closed. This is *the medial operation mode* and it is only observed in a narrow zone, defined by the change of the load resistor value which is however not immediate to no-load.

At modes, which are very close to no-load the third case is observed. The commutations in the rectifier now complete after the ones in the inverter, i.e. the rectifier diodes start conducting after the conduction beginning of the corresponding inverter's freewheeling diodes. This mode is *the boundary operation mode* with respect to no-load.

## 3. Analysis of the converter

In order to obtain general results, it is necessary to normalize all quantities characterizing the converter's state. The following quantities are included into relative units:

$x = U'_C = u_C/U_d$    - Voltage of the capacitor $C$;

$y = I' = \dfrac{i}{U_d/Z_0}$    - Current in the resonant circuit;

$U'_0 = kU_0/U_d$    - Output voltage;

$$I'_0 = \frac{I_0/k}{U_d/Z_0}$$   - Output current;

$U'_{Cm} = U_{Cm}/U_d$  - Maximum voltage of the capacitor $C$;

$v = \omega/\omega_0$       - Distraction of the resonant circuit,

where $\omega$ is the operating frequency and $\omega_0 = 1/\sqrt{LC}$ and $Z_0 = \sqrt{L/C}$ are the resonant frequency and the characteristic impedance of the resonant circuit $L$-$C$ correspondingly.

### 3.1 Analysis at the main operation mode of the converter

Considering the influence of the capacitors $C_S$ and $C_0$, the main operation mode of the converter can be divided into eight consecutive intervals, whose equivalent circuits are shown in fig. 2. By the trajectory of the depicting point in the state plane $(x = U'_C; y = I')$, shown in fig. 3, the converter's work is also illustrated, as well as by the waveform diagrams in fig.4.

The following four centers of circle arcs, constituting the trajectory of the depicting point, correspond to the respective intervals of conduction by the transistors and freewheeling diodes in the inverter: interval 1: $Q_1/Q_3$ - $(1-U'_0; 0)$; interval 3: $D_2/D_4$ - $(-1-U'_0; 0)$;

interval 5: $Q_2/Q_4$ - $(-1+U'_0; 0)$; interval 7: $D_1/D_3$ - $(1+U'_0; 0)$.

The intervals 2 and 6 correspond to the commutations in the inverter. The capacitors $C$ and $C_S$ then are connected in series and the sinusoidal quantities have angular frequency of $\omega'_0 = 1/\sqrt{LC_{E1}}$ where $C_{E1} = CC_S/(C+C_S)$. For the time intervals 2 and 6 the input current $i_d$ is equal to zero. These pauses in the form of the input current $i_d$ (fig. 4) are the cause for increasing the maximum current value through the transistors but they do not influence the form of the output characteristics of the converter.

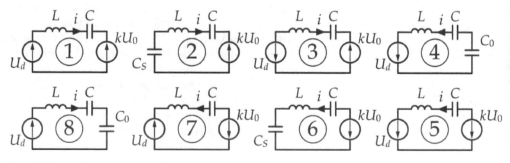

Fig. 2. Equivalent circuits at the main operation mode of the converter.

The intervals 4 and 8 correspond to the commutations in the rectifier. The capacitors $C$ and $C_0$ are then connected in series and the sinusoidal quantities have angular frequency of $\omega''_0 = 1/\sqrt{LC_{E2}}$ where $C_{E2} = CC_0/(C+C_0)$. For the time intervals 4 and 8, the output current $i_0$ is equal to zero. Pauses occur in the form of the output current $i_0$, decreasing its average value by $\Delta I_0$ (fig. 4) and essentially influence the form of the output characteristics of the converter.

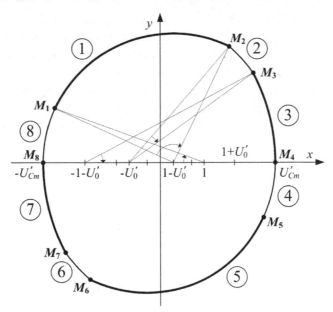

Fig. 3. Trajectory of the depicting point at the main mode of the converter operation.

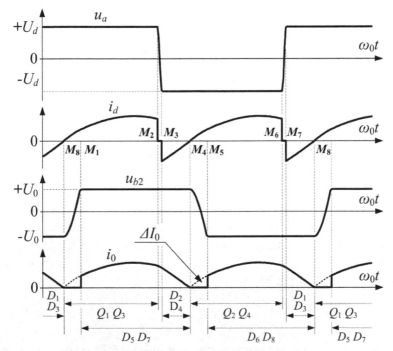

Fig. 4. Waveforms of the voltages and currents at the main operation mode of the converter

It has been proved in (Cheron, 1989; Bankov, 2009) that in the state plane (fig. 3) the points, corresponding to the beginning (p.$M_2$) and the end (p.$M_3$) of the commutation in the inverter belong to the same arc with its centre in point $(-U_0';0)$. It can be proved the same way that the points, corresponding to the beginning (p.$M_8$) and the end (p.$M_1$) of the commutation in the rectifier belong to an arc with its centre in point $(1;0)$. It is important to note that only the end points are of importance on these arcs. The central angles of these arcs do not matter either, because as during the commutations in the inverter and rectifier the electric quantities change correspondingly with angular frequencies $\omega_0'$ and $\omega_0''$, not with $\omega_0$.

The following designations are made:

$$a_1 = C_S/C \qquad n_1 = \sqrt{(a_1+1)/a_1} \tag{1}$$

$$a_2 = C_0/C \qquad n_2 = \sqrt{(a_2+1)/a_2} \tag{2}$$

$$n_3 = \sqrt{1+1/a_1+1/a_2} \tag{3}$$

For the state plane shown in fig, 3 the following dependencies are valid:

$$(x_1-1+U_0')^2 + y_1^2 = (x_2-1+U_0')^2 + y_2^2 \tag{4}$$

$$(x_2+U_0')^2 + y_2^2 = (x_3+U_0')^2 + y_3^2 \tag{5}$$

$$(x_3+1+U_0')^2 + y_3^2 = (x_4+1+U_0')^2 + y_4^2 \tag{6}$$

$$(x_4+1)^2 + y_4^2 = (x_5+1)^2 + y_5^2 \tag{7}$$

From the existing symmetry with respect to the origin of the coordinate system of the state plane it follows:

$$x_5 = -x_1 \tag{8}$$

$$y_5 = -y_1 \tag{9}$$

During the commutations in the inverter and rectifier, the voltages of the capacitors $C_S$ and $C_0$ change correspondingly by the values $2U_d$ and $2kU_0$, and the voltage of the commutating capacitor $C$ changes respectively by the values $2a_1U_d$ and $2a_2kU_0$. Consequently:

$$x_3 = x_2 + 2a_1 \tag{10}$$

$$x_5 = x_4 - 2a_2U_0' \tag{11}$$

The equations (4)÷(11) allow for calculating the coordinates of the points $M_1$÷$M_4$ in the state plane, which are the starting values of the current through the inductor L and the voltage of the commutating capacitor C in relative units for each interval of converter operation. The expressions for the coordinates are in function of $U'_0$, $U'_{Cm}$, $a_1$ and $a_2$ :

$$x_1 = -U'_{Cm} + 2a_2U'_0 \tag{12}$$

$$y_1 = \sqrt{4a_2U'_0\left(U'_{Cm} - a_2U'_0 + 1\right)} \tag{13}$$

$$x_2 = U'_0U'_{Cm} - a_2U'^2_0 - a_1 \tag{14}$$

$$y_2 = \sqrt{\begin{array}{l}\left(-U'_{Cm} + 2a_2U'_0 - U'_0U'_{Cm} + a_2U'^2_0 + a_1\right)\cdot \\ \cdot\left(-U'_{Cm} + 2a_2U'_0 + U'_0U'_{Cm} - a_2U'^2_0 - a_1 + 2U'_0 - 2\right) + \\ +4a_2U'_0\left(U'_{Cm} - a_2U'_0 + 1\right)\end{array}} \tag{15}$$

$$x_3 = U'_0U'_{Cm} - a_2U'^2_0 + a_1 \tag{16}$$

$$y_3 = \sqrt{\begin{array}{l}\left(U'_{Cm} - U'_0U'_{Cm} + a_2U'^2_0 - a_1\right)\cdot \\ \cdot\left(U'_{Cm} + U'_0U'_{Cm} - a_2U'^2_0 + a_1 + 2U'_0 + 2\right)\end{array}} \tag{17}$$

$$x_4 = U'_{Cm} \tag{18}$$

$$y_4 = 0 \tag{19}$$

For converters with only two reactive elements (L and C) in the resonant circuit the expression for its output current $I'_0$ is known from (Al Haddad et al., 1986; Cheron, 1989):

$$I'_0 = 2vU'_{C1m}/\pi \tag{20}$$

The LCC converter under consideration has three reactive elements in its resonant circuit (L, C и $C_0$). From fig.4 it can be seen that its output current $I'_0$ decreases by the value $\Delta I'_0 = 2a_2vU'_0/\pi$ :

$$I'_0 = 2vU'_{Cm}/\pi - \Delta I'_0 = 2v\left(U'_{Cm} - a_2U'_0\right)/\pi \tag{21}$$

The following equation is known:

$$\frac{\pi}{v} = \omega_0\left(t_1 + t_2 + t_3 + t_4\right), \tag{22}$$

where the times $t_1 \div t_4$ represent the durations of the different stages – from 1 to 4.
For the times of the four intervals at the main mode of operation of the converter within a half-cycle the following equations hold:

$$t_1 = \frac{1}{\omega_0}\left(\operatorname{arctg}\frac{y_2}{-x_2 + 1 - U'_0} - \operatorname{arctg}\frac{y_1}{-x_1 + 1 - U'_0}\right) \tag{23a}$$

at $\qquad x_2 \le 1 - U_0'$ and $x_1 \le 1 - U_0'$

$$t_1 = \frac{1}{\omega_0}\left(\pi - arctg\frac{y_2}{x_2 - 1 + U_0'} - arctg\frac{y_1}{-x_1 + 1 - U_0'}\right) \tag{23b}$$

at $\qquad x_2 \ge 1 - U_0'$ and $x_1 \le 1 - U_0'$

$$t_1 = \frac{1}{\omega_0}\left(arctg\frac{y_2}{-x_2 + 1 - U_0'} + arctg\frac{y_1}{x_1 - 1 + U_0'}\right) \tag{23c}$$

at $\qquad x_2 \ge 1 - U_0'$ and $x_1 \ge 1 - U_0'$

$$t_2 = \frac{1}{n_1\omega_0}\left(arctg\frac{n_1 y_2}{x_2 - 1 + U_0'} - arctg\frac{n_1 y_3}{x_3 + 1 + U_0'}\right) \tag{24a}$$

at $\qquad x_2 \ge 1 - U_0''$

$$t_2 = \frac{1}{n_1\omega_0}\left(arctg\frac{n_1 y_2}{x_2 - 1 + U_0'} - arctg\frac{n_1 y_3}{x_3 + 1 + U_0'}\right) \tag{24b}$$

at $\qquad x_2 \le 1 - U_0''$

$$t_3 = \frac{1}{\omega_0}arctg\frac{y_3}{x_3 + 1 + U_0'} \tag{25}$$

$$t_4 = \frac{1}{n_2\omega_0}\left(arctg\frac{n_2 y_1}{-x_1 + 1 - U_0'}\right) \tag{26a}$$

at $\qquad x_1 \le 1 - U_0'$

$$t_4 = \frac{1}{n_2\omega_0}\left(\pi - arctg\frac{n_2 y_1}{x_1 - 1 + U_0'}\right) \tag{26b}$$

at $\qquad x_1 \ge 1 - U_0'$

It should be taken into consideration that for stages 1 and 3 the electric quantities change with angular frequency $\omega_0$, while for stages 2 and 4 – the angular frequencies are respectively $\omega_0' = n_1\omega_0$ and $\omega_0'' = n_2\omega_0$.

### 3.2 Analysis at the boundary operation mode of the converter
At this mode, the operation of the converter for a cycle can be divided into eight consecutive stages (intervals), whose equivalent circuits are shown in fig. 5. It makes impression that the sinusoidal quantities in the different equivalent circuits have three different angular frequencies:

$\omega_0 = 1/\sqrt{LC}$ for stages 4 and 8;

$\omega_0' = 1/\sqrt{LC_{E2}}$, where $C_{E2} = CC_0/(C+C_0)$, for stages 1, 3, 5 and 7;

$\omega_0'' = 1/\sqrt{LC_{E3}}$, where $C_{E3} = CC_SC_0/(CC_S + CC_0 + C_SC_0)$, for stages 2 and 6.

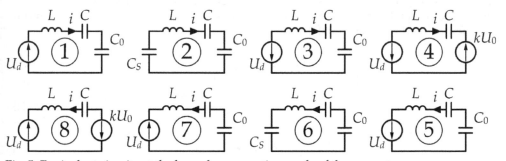

Fig. 5. Equivalent circuits at the boundary operation mode of the converter

In this case the representation in the state plane becomes complex and requires the use of two state planes (fig.6). One of them is $(x = U_C'; y = I')$ and it is used for presenting stages 4 and 8, the other is $(x^0; y^0)$, where:

$$x^0 = u_{C_{E2}}/U_d ; \qquad y^0 = \frac{i}{U_d/\sqrt{L/C_{E2}}} .$$

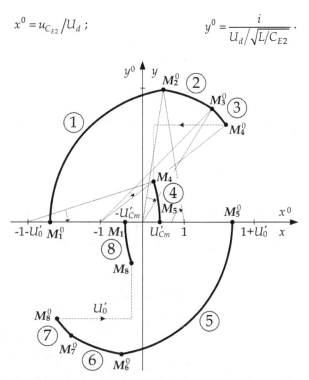

Fig. 6. Trajectory of the depicting point at the boundary mode of operation of the converter

Stages 2 and 6 correspond to the commutations in the inverter.

The commutations in the rectifier begin in p. $M_1^0$ or p. $M_5^0$ and end in p. $M_4^0$ or p. $M_8^0$, comprising stages 1,2 and 3 or 5,6 and 7.

The transistors conduct for the time of stages 1 and 5, the freewheeling diodes – for the time of stages 3, 4, 7 and 8, and the rectifier diodes – for the time of stages 4 and 8.

The following equations are obtained in correspondence with the trajectory of the depicting point for this mode of operation (fig.6):

$$\left(1-x_1^0\right)^2 = \left(1-x_2^0\right)^2 + \left(y_2^0\right)^2 \tag{27}$$

$$\left(x_2^0\right)^2 + \left(y_2^0\right)^2 = \left(x_3^0\right)^2 + \left(y_3^0\right)^2 \tag{28}$$

$$\left(x_3^0+1\right)^2 + \left(y_3^0\right)^2 = \left(x_4^0+1\right)^2 + \left(y_4^0\right)^2 \tag{29}$$

$$\left(x_4+1+U_0'\right)^2 + \left(y_4\right)^2 = \left(1+U_0'-x_1\right)^2 \tag{30}$$

During the commutations in the inverter, the voltage of the capacitor $C_{E2}$ changes by the value $2U_dC_S/C_{E2}$ and consequently:

$$x_3^0 = x_2^0 + 2a_1 n_2^2 \tag{31}$$

The same way during the commutation in the rectifier, the voltage of the capacitor $C_{E2}$ changes by the value $2kU_0C_0/C_{E2}$ and consequently:

$$x_4^0 = x_1^0 + 2U_0'\left(1+a_2\right) \tag{32}$$

From the principle of continuity of the current through the inductor $L$ and of the voltage in the capacitor $C$ it follows:

$$x_4^0 = x_4 + U_0' \tag{33}$$

$$y_4^0 = n_2 y_4 \tag{34}$$

where $(x_i;y_i)$ and $\left(x^0_i;y^0_i\right)$ are the coordinates of $M_i$ and $M_i^0$ respectively.

The equations (27)÷(34) allow for defining the coordinates of the points $M_1^0 \div M_4^0$ in the state plane:

$$x_1^0 = -U_{Cm}' - U_0' \tag{35}$$

$$y_1^0 = 0 \tag{36}$$

$$x_2^0 = \frac{\left(U_{Cm}' - a_2 U_0'\right)\left(1+U_0'+a_2 U_0'\right)}{a_2} - a_1 n_2^2 \tag{37}$$

$$y_2^0 = \sqrt{\left(U_{Cm}' + 1 + U_o'\right)^2 - \left(1 - x_2^0\right)^2} \tag{38}$$

$$x_3^0 = \frac{\left(U_{Cm}' - a_2 U_0'\right)\left(1 + U_0' + a_2 U_0'\right)}{a_2} + a_1 n_2^2 \tag{39}$$

$$y_3^0 = \sqrt{\left(x_2^0\right)^2 + \left(y_2^0\right)^2 - \left(x_3^0\right)^2} \tag{40}$$

$$x_4^0 = -U_{Cm}' + U_0'\left(1 + 2a_2\right) \tag{41}$$

$$y_4^0 = 2\sqrt{\left(a_2 U_0' + 1 + U_0'\right)\left(a_2 U_0' - U_{Cm}'\right)} \tag{42}$$

At the boundary operation mode the output current $I_0'$ is defined by expression (21) again, where $t_1 \div t_4$ represent the times of the different stages – from 1 to 4.

For the times of the four intervals at the boundary operation mode of the converter within a half-cycle the following equations hold:

$$t_1 = \frac{1}{n_2 \omega_0}\left(arctg\frac{y_2^0}{1 - x_2^0}\right) \qquad \text{at } x_2^0 \le 1 \tag{43a}$$

$$t_1 = \frac{1}{n_2 \omega_0}\left(\pi - arctg\frac{y_2^0}{x_2^0 - 1}\right) \qquad \text{at } x_2^0 \ge 1 \tag{43b}$$

$$t_2 = \frac{1}{n_3 \omega_0}\left(\pi - arctg\frac{n_3 y_2^0}{n_2\left(1 - x_2^0\right)} - arctg\frac{n_3 y_3^0}{n_2\left(1 + x_3^0\right)}\right) \qquad \text{at } x_2^0 \le 1 \tag{44a}$$

$$t_2 = \frac{1}{n_3 \omega_0}\left(arctg\frac{n_3 y_2^0}{n_2\left(x_2^0 - 1\right)} - arctg\frac{n_3 y_3^0}{n_2\left(1 + x_3^0\right)}\right) \qquad \text{at } x_2^0 \ge 1 \tag{44b}$$

$$t_3 = \frac{1}{n_2 \omega_0}\left(arctg\frac{y_3^0}{1 + x_3^0} - arctg\frac{y_4^0}{1 + x_4^0}\right) \tag{45}$$

$$t_4 = \frac{1}{\omega_0}\left(arctg\frac{y_4}{x_4 + 1 + U_0'}\right) \tag{46}$$

It should be taken into consideration that for stages 1 and 3 the electric quantities change by angular frequency $\omega_0'' = n_2 \omega_0$, while for stages 2 and 4 the angular frequencies are correspondingly $\omega_0''' = n_3 \omega_0$ and $\omega_0$.

## 4. Output characteristics and boundary curves

On the basis of the analysis results, equations for the output characteristics are obtained individually for both the main and the boundary modes of the converter operation. Besides, expressions for the boundary curves of the separate modes are also derived.

### 4.1 Output characteristics and boundary curves at the main operation mode

From equation (21) $U'_{Cm}$ is expressed in function of $U'_0$, $I'_0$, v and $a_2$

$$U'_{Cm} = \frac{\pi}{2v}I'_0 + a_2 U'_0 \qquad (47)$$

By means of expression (47) $U'_{Cm}$ is eliminated from the equations (12)÷(18). After consecutive substitution of expressions (12)÷(18) in equations (23)÷(26) as well as of expressions (23)÷(26) in equation (22), an expression of the kind $U'_0 = f(I'_0, v, a_1, a_2)$ is obtained. Its solution provides with the possibility to build the output characteristics of the converter in relative units at the main mode of operation and at regulation by means of changing the operating frequency. The output characteristics of the converter respectively for v=1.2; 1.3; 1.4; 1.5; 1.8; 2.5; 3.0; 3.3165; 3,6 and $a_1$=0.1; $a_2$=0.2 as well as for v=1.2; 1.3; 1.4; 1.5; 1.6; 1.8 and $a_1$=0.1; $a_2$=1.0 are shown in fig.7-a and 7-b.

The comparison of these characteristics to the known ones from (Al Haddad et al., 1986; Cheron, 1989) shows the entire influence of the capacitors $C_S$ и $C_0$. It can be seen that the output characteristics become more vertical and the converter can be regarded to a great extent as a source of current, stable at operation even at short circuit. Besides, an area of operation is noticeable, in which $U'_0 > 1$.

At the main operation mode the commutations in the rectifier (stages 4 and 8) must always end before the commutations in the inverter have started. This is guaranteed if the following condition is fulfilled:

$$x_1 \leq x_2 \qquad (48)$$

In order to enable natural switching of the controllable switches at zero voltage (ZVS), the commutations in the inverter (stages 2 and 6) should always end before the current in the resonant circuit becomes zero. This guaranteed if the following condition is fulfilled:

$$x_3 \leq U'_{Cm} \qquad (49)$$

If the condition (49) is not fulfilled, then switching a pair of controllable switches off does not lead to natural switching the other pair of controllable switches on at zero voltage and then the converter stops working. It should be emphasized that these commutation mistakes do not lead to emergency modes and they are not dangerous to the converter. When it „misses", all the semiconductor switches stop conducting and the converter just stops working. This is one of the big advantages of the resonant converters working at frequencies higher than the resonant one.

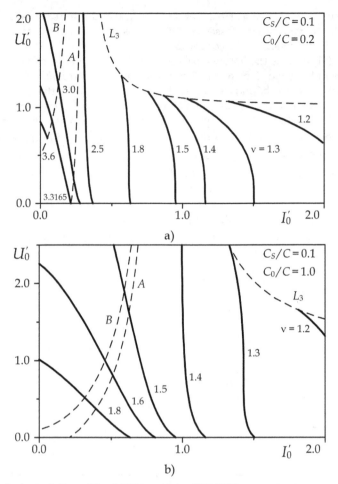

Fig. 7. Output characteristics of the LCC transistor DC/DC converter

Equations (12), (14) and (47) are substituted in condition (48), while equations (16) and (47) are substituted in condition (49). Then the inequalities (48) and (49) obtain the form:

$$I_0' \geq \frac{2v}{\pi} \cdot \frac{a_1 + a_2 U_0'}{1 + U_0'} \tag{50}$$

$$I_0' \geq \frac{2v}{\pi} \cdot \frac{a_1 - a_2 U_0'}{1 - U_0'} \tag{51}$$

Inequalities (50) и (51) enable with the possibility to draw the boundary curve $A$ between the main and the medial modes of operation of the converter, as well as the border of the natural commutation – curve $L_3$ (fig. 7-a) or curve $L_4$ (fig. 7-b) in the plane of the output characteristics. It can be seen that the area of the main operation mode of the converter is limited within the boundary curves $A$ and $L_3$ or $L_4$. The bigger the capacity of the capacitors

$C_S$ and $C_0$, the smaller this area is. However, the increase of the snubber capacitors leads to a decrease in the commutation losses in the transistors as well as to limiting the electromagnetic interferences in the converter.

The expression (51) defines the borders, beyond which the converter stops working because of the breakage in the conditions for natural switching the controllable switches on at zero voltage (ZVS). Exemplary boundary curves have been drawn in the plane of the output characteristics (fig.8) at $a_1=0.10$. Four values have been chosen for the other parameter: $a_2 = 0.05; 0.1; 0.2$ and $1.0$. When the capacity of the capacitor $C_0$ is smaller or equal to that of the snubber capacitors $C_S$ $(a_1 \geq a_2)$, then the converter is fit for work in the area between the curve $L_1$ or $L_2$ and the x-axis (the abscissa). Only the main operation mode of the converter is possible in this area. The increase in the load resistance or in the operating frequency leads to stopping the operation of the converter before it has accomplished a transition towards the medial and the boundary modes of work.

When $C_0$ has a higher value than the value of $C_S$ $(a_1 < a_2)$ then the boundary curve of the area of converter operation with ZVS is displaced upward (curve $L_3$ or $L_4$). It is possible now to achieve even a no-load mode.

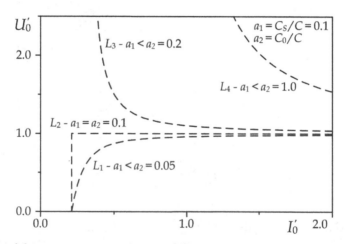

Fig. 8. Borders of the converter operation capability

## 4.2 Output characteristics and boundary curves at the boundary operation mode

Applying expression (47) for equations (35)÷(42) $U'_{Cm}$ is eliminated. After that, by a consecutive substitution of expressions (35)÷(42) in equations (43)÷(45) as well as of expressions (43)÷(46) in equation (22), a dependence of the kind $U'_0 = f(I'_0, v, a_1, a_2)$ is obtained. Its solving enables with a possibility to build the outer (output) characteristics of the converter in relative units at the boundary operation mode under consideration and at regulation by changing the operating frequency. Such characteristics are shown in fig. 7-a for v = 3.0; 3.3165; 3.6 and $a_1=0.1$; $a_2=0.2$ and in fig. 7-b for v =1.5; 1.6; 1.8 and $a_1=0.1$; $a_2=1.0$.

At the boundary operation mode, the diodes of the rectifier have to start conducting after opening the freewheeling diodes of the inverter. This is guaranteed if the following condition is fulfilled:

$$x_4^0 \geq x_3^0 \tag{52}$$

After substitution of equations (39), (41) и (47) in the inequality (52), the mentioned above condition obtains the form:

$$I_0' \leq \frac{2\nu}{\pi} \cdot \frac{a_2 U_0' - a_1}{1 + U_0'} \tag{53}$$

Condition (53) gives the possibility to define the area of the boundary operation mode of the converter in the plane of the output characteristics (fig. 7-a and fig. 7-b). It is limited between the y-axis (the ordinate) and the boundary curve $B$. It can be seen that the converter stays absolutely fit for work at high-Ohm loads, including at a no-load mode. It is due mainly to the capacitor $C_0$. With the increase in its capacity (increase of $a_2$) the area of the boundary operation mode can also be increased.

## 5. Medial operation mode of the converter

At this operation mode, the diodes of the rectifier start conducting during the commutation in the inverter. The equivalent circuits, corresponding to this mode for a cycle, are shown in fig.9. In this case, the sinusoidal quantities have four different angular frequencies:

$\omega_0 = 1/\sqrt{LC}$ for stages 4 and 8;

$\omega_0' = 1/\sqrt{LC_{E1}}$ , where $C_{E1} = CC_S/(C + C_S)$ , for stages 3 and 7;

$\omega_0'' = 1/\sqrt{LC_{E2}}$ , where $C_{E2} = CC_0/(C + C_0)$ , for stages 1 and 5;

$\omega_0''' = 1/\sqrt{LC_{E3}}$ , where $C_{E3} = CC_S C_0/(CC_S + CC_0 + C_S C_0)$ , for stages 2 and 6.

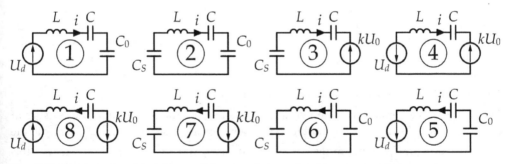

Fig. 9. Equivalent circuits at the medial operation mode of the converter

Therefore, the analysis of the medial operation mode is considerably more complex. The area in the plane of the output characteristics, within which this mode appears, however, is completely defined by the boundary curves $A$ and $B$ for the main and the boundary modes respectively. Having in mind the monotonous character of the output characteristics for the other two modes, their building for the mode under consideration is possible through linear interpolation. It is shown in fig. 7-a for $\nu = 3.0$; 3.3165 as well as in fig. 7-b for $\nu=1.5$; 1.6; 1.8. The larger area of this mode corresponds to the higher capacity of the snubber capacitors $C_S$ and the smaller capacity of the capacitor $C_0$.

## 6. Methododlogy for designing the converter

During the process of designing the $LCC$ resonant $DC/DC$ converter under consideration, the following parameters are usually predetermined: power in the load $P_0$, output voltage $U_0$ and operating frequency $f$. Very often, the value of the power supply voltage $U_d$ is predefined and the desired output voltage is obtained by adding a matching transformer. The design of the converter could be carried out in the following order:

1. Choice of the frequency distraction ν

For converters, operating at frequencies higher than the resonant one, the frequency distraction is usually chosen in the interval ν =1.1÷1.3. It should be noted here that if a higher value of ν is chosen, the exchange of reactive energy between the resonant circuit and the energy source increases, i.e., the current load on the elements in the circuit increases while the stability of the output characteristics decreases. It is due to the increase in the impedance of the resonant circuit of the converter.

2. Choice of the parameter $a_2 = C_0/C$

The choice of this parameter is a compromise to a great extent. On the one hand, to avoid the considerable change of the operating frequency during the operation of the converter from nominal load to no-load mode, the capacitor $C_0$ (respectively the parameter $a_2$) should be big enough. On the other hand, with the increase of $C_0$ the current in the resonant circuit increases and the efficiency of the converter decreases. It should be noted that a significant increase in the current load on the elements in the scheme occurs at $a_2 > 1$ (Malesani et al., 1995).

3. Choice of the parameter $a_1 = C_S/C$

The parameter $a_1$ is usually chosen in the interval $a_1$ = 0.02÷0.20. The higher the value of $a_1$ (the bigger the capacity of the damping capacitors), the smaller the area of natural commutation of the transistors in the plane of the output characteristics is. However, the increase in the capacity of the snubber capacitors leads to a decrease in the commutation losses and limitation of the electromagnetic interferences in the converter.

4. Choice of the coordinates of a nominal operating point

The values of the parameters $a_1$ and $a_2$ fully define the form of the output characteristics of the converter. The nominal operating point with coordinates $I'_0$ and $U'_0$ lies on the characteristic, corresponding to the chosen frequency distraction ν. It is desirable that this point is close to the boundary of natural commutation so that the inverter could operate at a high power factor (minimal exchange of reactive energy). If the converter operates with sharp changes in the load and control frequency, however, it is then preferable to choose the operating point in an area, relatively distant from the border of natural commutation. Thus, automatic switching the converter off is avoided – a function, integrated in the control drivers of power transistors.

5. Defining the transformation ratio of the matching transformer

The transformation ratio is defined, so that the required output voltage of the converter at minimal power supply voltage and at maximum load is guaranteed:

$$k \geq \frac{U_0' \cdot U_{d\min}}{U_0}, \tag{54}$$

where $U_{d\min}$ is the minimal permissible value of the input voltage $U_d$.

6. Calculating the parameters of the resonant circuit

The values of the elements in the resonant circuit $L$ and $C$ are defined by the expressions related to the frequency distraction and the output current in relative units:

$$v = \omega/\omega_0 = 2\pi f \sqrt{LC} \qquad ; \qquad I_0' = \frac{I_0/k}{U_d/\sqrt{L/C}} = \frac{P_0/k}{U_d U_0/\sqrt{L/C}} \tag{55}$$

Solving the upper system of equations, it is obtained:

$$L = \frac{kv U_d U_0 I_0'}{2\pi f P_0} \qquad ; \qquad C = \frac{v P_0}{2\pi f k U_d U_0 I_0'} \tag{56}$$

## 7. Experimental investigations

For the purposes of the investigation, a laboratory prototype of the LCC resonant converter under consideration was designed and made without a matching transformer and with the following parameters: power supply voltage $U_d = 500$ V; output power $P_0 = 2.6$ кW, output voltage $U_0 = 500$ V; operating frequency and frequency distraction at nominal load $f = 50$ kHz and $v = 1.3$; $a_1 = 0.035$; $a_2 = 1$; coordinates of the nominal operating point - $I_0' = 1.43$ and $U_0' = 1$. The following values of the elements in the resonant circuit were obtained with the above parameters: $L = 570$ µH; $C = C_0 = 30$ nF. The controllable switches of the inverter were IGBT transistors with built-in backward diodes of the type IRG4PH40UD, while the diodes of the rectifier were of the type BYT12PI. Snubber capacitors $C_1 \div C_4$ with capacity of 1 nF were connected in parallel to the transistors. Each transistor possessed an individual driver control circuit. This driver supplied control voltage to the gate of the corresponding transistor, if there was a control signal at the input of the individual driver circuit and if the collector-emitter voltage of the transistor was practically zero (ZVC commutation).

Experimental investigation was carried out during converter operation at frequencies $f = 50$ kHz ($v = 1.3$) and $f = 61.54$ kHz ($v = 1.6$). The dotted curve in fig.10 shows the theoretical output characteristics, while the continuous curve shows the output characteristics, obtained in result of the experiments.

A good match between the theoretical results and the ones from the experimental investigation can be noted. The small differences between them are mostly due to the losses in the semiconductor switches in their open state and the active losses in the elements of the resonant circuit.

Oscillograms, illustrating respectively the main and the boundary operation modes of the converter are shown in fig. 11 and fig. 12. These modes are obtained at a stable operating frequency $f = 61.54$ kHz ($v = 1.6$) and at certain change of the load resistor. In the oscillograms the following quantities in various combinations are shown: output voltage ($u_a$) and output current ($i$) of the inverter, input voltage ($u_b$) and output current ($i_0$) of the rectifier.

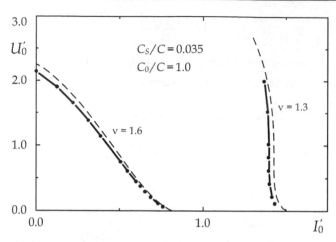

Fig. 10. Experimental output characteristics of the converter.

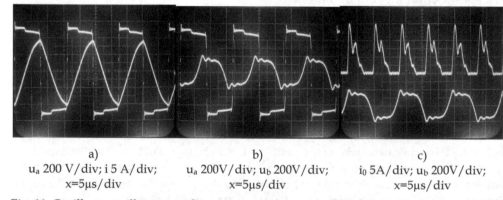

a)
$u_a$ 200 V/div; i 5 A/div;
x=5μs/div

b)
$u_a$ 200V/div; $u_b$ 200V/div;
x=5μs/div

c)
$i_0$ 5A/div; $u_b$ 200V/div;
x=5μs/div

Fig. 11. Oscillograms illustrating the main operation mode of the converter

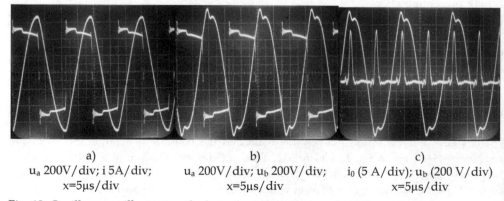

a)
$u_a$ 200V/div; i 5A/div;
x=5μs/div

b)
$u_a$ 200V/div; $u_b$ 200V/div;
x=5μs/div

c)
$i_0$ (5 A/div); $u_b$ (200 V/div)
x=5μs/div

Fig. 12. Oscillograms illustrating the boundary operation mode of the converter

From fig. 11-b and fig. 12-b the difference between the main and the boundary operation mode of the converter can be seen. In the first case, the commutations in the rectifier (the process of recharging the capacitor $C_0$) end before the commutations in the inverter (the process of recharging the capacitor $C_S$). In the second case, the commutations in the rectifier complete after the ones in the inverter. In both cases during the commutations in the rectifier, all of its diodes are closed and the output current $i_0$ is equal to zero (fig. 11-c and fig. 12-c).

Fig. 12-b confirms the fact that at certain conditions the output voltage can become higher than the power supply voltage without using a matching transformer.

At no-load mode, the converter operation is shown in fig. 13. In this case, the output voltage is more than two times higher than the power supply one.

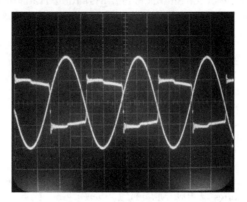

$u_a$ 500V/div; $u_b$ 500V/div; x=5µs/div

Fig. 13. Oscillograms, illustrating no-load mode of the converter

## 8. Conclusions

The operation of an LCC transistor resonant DC/DC converter with a capacitive output filter and working above the resonant frequency has been investigated, taking into account the influence of snubber capacitors and a matching transformer. The particular operation modes of the converter have been considered, and the conditions under which they are obtained have been described. The output characteristics for all operation modes of the converter have been built including at regulation by means of changing the operating frequency. The boundary curves between the different operation modes of the converter as well as the area of natural commutation of the controllable switches have been shown in the plane of the output characteristics. Results from investigations carried out by means of a laboratory prototype of the converter have been obtained and these results confirm the ones from the analysis.

The theoretical investigations show that the conditions for ZVS can be kept the same for high-Ohm loads and the converter can stay fit for work even at a no-load mode. For the purpose, it is necessary to have the natural capacity of the matching transformer bigger than the one of the snubber capacitors.

The output characteristics show that in the zone of small loads the value of the normalized output voltage increases to reach a value higher than unit what is characteristic for converters with controllable rectifying. This can be explained by the similar mechanism of the rectifier operation in the investigated converter.

The results from the investigation can be used for more precise designing of LCC converters used as power supplies for electric arc welding aggregates, powerful lasers, luminescent lamps etc.

## 9. References

Al Haddad, K., Cheron, Y., Foch, H. & Rajagopalan, V. (1986). Static and dynamic analysis of a series resonant converter operating above its resonant frequency, *Proceedings of SATECH'86*, pp.55-68, Boston, USA.

Bankov, N. (2009) Influence of the Snubbers and Matching Transformer over the Work of a Transistor Resonant DC/DC Converter. *Elektrotehnika&Elektronika (Sofia, Bulgaria)*, Vol. 44, No. 7-8, pp. 62-68, ISSN 0861-4717.

Cheron, Y., Foch, H. & Salesses, J. (1985). Study of resonant converter using power transistors in a 25-kW X-Rays tube power supply. *IEEE Power Electronics Specialists Conference*, ESA Proceedings, 1985, pp. 295-306.

Cheron, Y. (1989). *La commutation douce dans la conversion statique de l'energie electrique*, Technique et Documentation, ISBN : 2-85206-530-4, Lavoisier, France.

Malesani, L., Mattavelli, P., Rossetto, L., Tenti, P., Marin, W. & Pollmann, A. (1995). Electronic Welder With High-Frequency Resonant Inverter. *IEEE Transactions on Industry Applications*, Vol. 31, No.2, (March/April 1995), pp. 273-279, ISSN: 0093-9994.

Jyothi, G. & Jaison, M. (2009). Electronic Welding Power Source with Hybrid Resonant Inverter, *Proceedings of 10th National Conference on Technological Trends (NCTT09)*, pp. 80-84, Kerala, India, 6-7 Nov 2009.

Liu, J., Sheng, L., Shi, J., Zhang, Z. & He, X. (2009). Design of High Voltage, High Power and High Frequency in LCC Resonant Converter. *Applied Power Electronics Conference and Exposition, APEC 2009. Twenty-Fourth Annual IEEE*, pp. 1034-1038, ISSN: 1048-2334, Washington, USA, 15-19 Feb. 2009.

Ivensky, G., Kats, A. & Ben-Yaakov, S. (1999). An RC load model of parallel and series-parallel resonant DC-DC converters with capacitive output filter. *IEEE Transactions on Power Electronics*, Vol. 14, No.3, (May 1999), pp. 515-521, ISSN: 0885-8993.

# Part 3

## Harmonic Distortion

# 6

# Improve Power Quality with High Power UPQC

Qing Fu, Guilong Ma and Shuhua Chen
*Sun Yat-sen University*
*China*

## 1. Introduction

An ideal AC power transmission is pure sinusoidal, both its voltage and its current. With the increasing production of modern industry, more and more power electronic equipments are used and cause serious current distortion because of open and close of power electronic devices. Harmonic, a measurement of distorted degree of voltage or current, reflects the deviation from sinusoidal wave. Another cause of harmonic is nonlinear loads such as Arc furnaces and transformers. The widely using of nonlinear load brings much harmonic current to transmission lines. The harmonic current passes through transmission lines and causes harmonic voltage exert on the loads in other place(Terciyanli et al. 2011). As a result, the loss of power transmission is increased and the safety of power grid is seriously weakened.

With the fast development of modern production, the harmonic in power grid become more and more serious and people pay more attention to how to eliminate harmonic(wen et al. 2010). Active Power Filter (APF) is a promising tool to cut down the influence of harmonics, shunt APF for harmonic current, series APF for harmonic voltage. Unified Power Quality Conditioner (UPQC), consisted of shunt APF and series APF, is effective to reduce both harmonic voltage and harmonic current. Now, UPQC is mainly used in low-voltage low-capacity applications. But with the development of power system, more and more high-power nonlinear loads are connected to higher voltage grid and the demand of high voltage and high capacity keeps being enlarged. The paper discussed a high power UPQC for high power nonlinear loads. In this UPQC, shunt APF uses a hybrid APF which includes a Passive Power Filter (PPF) and an APF. Shunt APF is connected to a series LC resonance circuit in grid fundamental frequency so as to make shunt APF in lower voltage and lower power. The series LC resonance circuit is connected to grid with a capacitor. DC linker of PPF is connected to DC link of APF. This type of UPQC is fit for high voltage high power application because the voltage and capacity of its active device is much lower than those of the whole UPQC. The paper discussed the principle and control method of this UPQC.

## 2. Fundamental knowledge

To show better about the principle and the theory about the high power UPQC, some fundamental knowledge about harmonic and harmonic elimination equipments are list below.

## 2.1 Series active power filter

In power system, voltage out from turbine is promising to be sinusoidal. So if there is no nonlinear load connects to power grid between generator and the nonlinear load in question, a shunt APF is enough to keep both the voltage and the current of transmission line sinusoidal because the transmission line is composed of linear components such as resistances, inductions and capacitors. But in modern power system, power is transmitted for a long distance before delivery to the nonlinear load and power is distributed to many nonlinear loads in many difference places along the transmission line. The transmission of harmonic current causes harmonic voltage in transmission lines which increases possibility of damage to some critical loads such as storage devices and some micromachining devices. Shunt APF can do little with the damage caused by harmonic voltage in transmission line. A series APF is installed between power source and critical load so as to insulate voltage harmonic from the critical load(Kim et al. 2004). It is also promising to eliminate damages to load caused by some other supply quality issues such as voltage sage, instant voltage interrupts, flicks and over voltage.

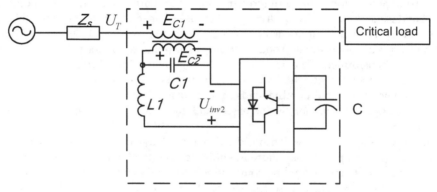

Fig. 1. Configuration of series APF

## 2.2 Shunt active power filter

The distortion of current not only brings serious loss of power transmission, but also endangers power grid and power equipments. Harmonic current increases the current flowed through transmission lines and as a result power transmission loss is increased and power grid has to take a risk of higher temperature which threatens the safety of power grid. Harmonic current in transformers will make them magnetic saturated and seriously heated. Much noise is generated because of harmonics in equipments. Besides, harmonics make some instruments indicate or display wrong values, and sometimes make they work wrong.

To eliminate harmonic current produced by nonlinear loads, a shunt Active Power Filter (APF) is expected to connect parallel to power grid(Ahmed et al. 2010). Shunt APF draws energy from power grid and makes it to be harmonic current that is equal to the harmonic current produced by nonlinear load so that harmonic current doesn't go to transmission line but goes between nonlinear load and APF. Usually an inverter is employed to realize this function.

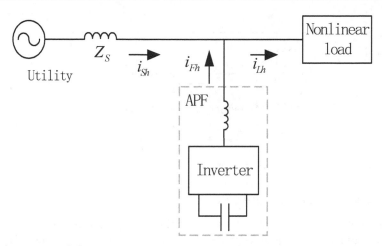

Fig. 2. Configuration of shunt APF

Fig.2 shows Configuration of shunt APF, where $Z_s$ is impedance of transmission line, $i_{sh}$ is harmonic current trough transmission line, $i_{Lh}$ is load harmonic current and $i_{Fh}$ is harmonic current from APF. APF employs an inverter to generator a harmonic current that always keeps equal to load harmonic current, that is:

$$i_{Fh} \equiv i_{Lh} \tag{1}$$

Then load harmonic current is intercepted by APF and will not pass through transmission line.

$$i_{sh} \equiv 0 \tag{2}$$

Usually a voltage source inverter which uses a high capacity capacitor to store energy in DC linker is used.

Under some conditions, nonlinear load not only produces harmonic current but also produces much more reactive current. In order to avoid reactive current going to transmission line, the shunt equipment needs to compensate also the reactive current. Passive Power Filter (PPF) is usually added to APF to compensate most of reactive current and a part of harmonic current so as to decrease the cost. This hybrid system of APF and PF is called Hybrid Active Power Filter (HAPF) (Wu et al. 2007). In HAPF, APF and PPF are connected in different forms and form many types of HAPF. Because of its low cost, HAPF attracts more and more eyes and has been developing very quickly.

### 2.3 UPQC: Combined shunt APF and series APF

Unified Power Quality Conditioner (UPQC) is composed of series APF and shunt APF(Yang & Ren, 2008). It not only protects the critical load from voltage quality problems but also eliminates the harmonic current produced by load. In UPQC, the series APF (usually called its series device) and shunt APF (usually called its shunt device) usually share the energy storage so as to simplify the structure and reduce the cost of UPQC.

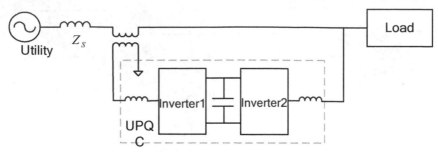

Fig. 3. Unified Power Quality Conditioner

## 3. An UPQC in high power application

In many mid-voltage or high-voltage applications, nonlinear load not only produces heavy harmonic current but also is sensitive to harmonic voltage. An UPQC combined a series APF and a HAPF is much suitable for these applications(Khadkikar et al.,2005). Fig.4 shows the detailed system configuration of the high power UPQC, where $e_{sa}$, $e_{sb}$ and $e_{sc}$ are three phase voltages of generator, $e_{ca}$, $e_{cb}$ and $e_{cc}$ are the voltages compensated by series APF, $I_s$ is utility current, $I_L$ is load current, $I_F$ is compensating current output from shunt device, $Z_s$ is impedance of transmission line, C is a big capacitor for DC linker.

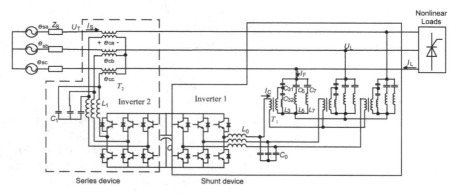

Fig. 4. Configuration of high power UPQC

The high power UPQC is composed of series device and shunt device. The series device is mainly for insulating the source voltage interference, adjusting loads voltage etc. The shunt device is mainly for eliminating harmonic current produced by nonlinear load. In series device, $L_1$ and $C_1$ make low-pass filter (LPF) to filter output voltage of Inverter 2 because power electronics devices in Inverter 2 open and close in high frequency and generate high frequency disturbances exerted on expected sinusoidal output voltage of Inverter 2. In series device, transformer $T_2$ not only insulates Inverter 2 from utility but also makes output voltage of Inverter 2 (after LPF) satisfy maximum utility harmonic voltage. In shunt device, $L_0$ and $C_0$ make a LPF to filter output voltage of Inverter 1. The shunt device and series device share the DC capacitor. The shunt device is consisted of an inverter and a PPF. PPF is consisted of 3 L-C resonance branches. One is consisted of $L_5$ and $C_5$ for 5th harmonic current elimination, the other is consisted of $L_7$ and $C_7$ for 7th harmonic current elimination, and the third is consisted of $L_3$, $C_{31}$, $C_{32}$ for 3rd harmonic current elimination. The resonance frequency of $L_3$ and $C_{32}$ is set to be the same as the frequency of fundamental component so that most of fundamental reactive current in this series resonance branch goes through $L_3$ and $C_{32}$ and little goes through inverter through transformer $T_1$. As a result Inverter 1 suffers little fundamental voltage which helps to cut down its cost and improve its safety. Transformer T1 connects Inverter 1 with the series fundamental resonant branch $L_3$ and $C_{32}$ to insulate them and fit the difference between maximum output voltage of Inverter 1 and maximum voltage that L3 and $C_{32}$ needed to generate the maximum compensating current. The 3rd, 5th, 7th harmonic currents can be eliminated by the 3 L-C resonance branches, and Inverter 1 can also inject harmonic current into utility to give a fine compensation to every order harmonic current except 3rd harmonic current.

### 3.1 Series device of high power UPQC
Series device of UPQC is mainly to filter utility voltage and adjust voltage exerted on load so as to eliminate harmonic current produced by utility harmonic voltage and provide load a good sinusoidal voltage(Brenna et al. 2009; Zhou et al. 2009).

Series device of high power UPQC has the same topology as series APF whose Configuration is shown in Fig.1. Fig.1shows the single phase equivalent circuit of the series device, where $Z_s$ is impedance of transmission line. The main circuit and control circuit of the active part are in the dashed box.

From the sigle-pahse system, the voltage of the transformer can be expressed as

$$E_{C2} = U_{inv2} \cdot \frac{Z_{C1}}{Z_{L1} + Z_{C1}} \tag{3}$$

Suppose $E_{C1} = n \cdot E_{C2}$, then the voltage of the Inverter 2 can be calculated as

$$U_{inv2} = E_{C2} \cdot \frac{Z_{L1} + Z_{C1}}{Z_{C1}}$$
$$= \frac{Z_{L1} + Z_{C1}}{nZ_{C1}} (U_T - U_L) \tag{4}$$

The voltage of Inverter 2 can be written at another way as

$$U_{inv2} = K_V \cdot U_{DC} \cdot B(s) \tag{5}$$

Where $K_V$ is amplitude ratio between $U_{inv2}$ and $U_{DC}$, $B(s)$ is phase shift between input control signal and output voltage of Inverter 2.

$$U_L = U_T - E_{C1}$$
$$= U_T - n \cdot K_V \cdot B(s) \cdot \frac{Z_{C1}}{Z_{L1} + Z_{C1}} \cdot U_{DC} \tag{6}$$
$$= U_T - K_{CL} U_{DC}$$

Where
$$K_{CL} = n \cdot K_V \cdot B(S) \cdot \frac{Z_{C1}}{Z_{L1} + Z_{C1}} \tag{7}$$

To make load voltage sinusoidal, load voltage $U_L$ is usually sampled for control. Control scheme for series device is:

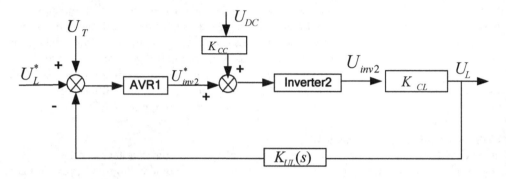

Fig. 5. Control scheme for series device of high power UPQC

Where AVR1 is automatic voltage regulator for $U_L$ control and AVR2 is for $U_C$ control. $U_{DC}$ is voltage of DC-linker. $K_{UC}(S)$ is transform function of detecting circuit of $U_C$ which is consisted of a proportion segment and a delay segment. $K_{UL}(S)$ is transform function of detecting circuit of $U_L$. $U^*_L$ is reference voltage for load voltage $U_L$, when a certain harmonic component is concerned, it is set to zero. AVR1 is automatic voltage regulator for $U_L$ and it can be divided to 3 parts, one is harmonic extraction, another is PI adjustor and the third is delay array. Control scheme of AVR1 is depicted in Fig.6. A selective harmonic extraction is adopted to extract the main order harmonics. Abc_dq0 is described as equation (8-10) for a certain k order harmonic and transformation dq0_abc is described as equation (11-13). LPF is low pass filter that only let DC component pass through.

$$U_d = \frac{2}{3}(V_a \sin(k\omega_0) + V_b \sin[k(\omega_0 - \frac{2\pi}{3})] + V_c \sin[k(\omega_0 + \frac{2\pi}{3})] \tag{8}$$

$$U_q = \frac{2}{3}(V_a \cos(k\omega_0) + V_b \cos[k(\omega_0 - \frac{2\pi}{3})] + V_c \cos[k(\omega_0 + \frac{2\pi}{3})] \tag{9}$$

$$U_0 = \frac{1}{3}(V_a + V_b + V_c) \tag{10}$$

$$V_a = U_d \sin(k\omega_0) + U_q \cos(k\omega_0) + U_0 \tag{11}$$

$$V_b = U_d \sin[k(\omega_0 - \frac{2\pi}{3})] + U_q \cos[k(\omega_0 - \frac{2\pi}{3})] + U_0 \tag{12}$$

$$V_c = U_d \sin[k(\omega_0 + \frac{2\pi}{3})] + U_q \cos[k(\omega_0 + \frac{2\pi}{3})] + U_0 \tag{13}$$

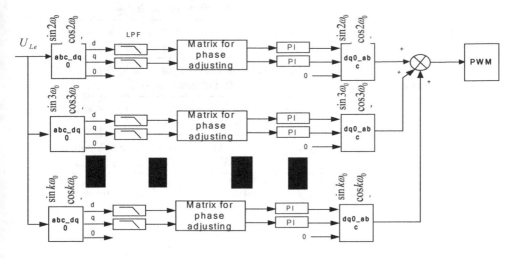

Fig. 6. Control scheme of AVR1

Because a delay will unavoidably happen during detecting and controlling, a matrix is used to adjust the phase shift of the certain order harmonic. The matrix is described as:

$$\begin{bmatrix} U_d' \\ U_q' \end{bmatrix} = \begin{bmatrix} \cos(k\omega_0 + \theta) & -\sin(k\omega_0 + \theta) \\ \sin(k\omega_0 + \theta) & \cos(k\omega_0 + \theta) \end{bmatrix} \begin{bmatrix} U_d \\ U_q \end{bmatrix} \tag{14}$$

Where $\theta$ is phase angle for delay.

To check the effect of series device of high power UPQC to harmonic voltage, with MATLAB, a 3-phase 10KV utility supplied to capacitors is set up. Suppose the initial load is a 3-phase capacitor group, a resister valued 0.2 ohm series with a capacitor valued 100uF in each phase. When t=0.04s, series device switches to run. Tab.1 shows the parameters of power source and series device. Comparing the main harmonic voltages and harmonic currents after series run with those before series run, we know that series device reduce much harmonic of load voltage and so load harmonic current is much reduced. Fig.7 shows waveform of load voltage before and after series device run. In Fig.8, the spectrums of load voltage are compared through FFT. Fig.9 shows load current waveform and Fig.11 shows the spectrums of load current before and after series device run. With transformer T2, fundamental voltage produced by Inverter 2 can be added to power source, so it can also compensate voltage sags. When it is concerned, $U_L^*$ in Fig.6 is set to be expected fundament component of source voltage. Fig.12 and Fig.13 shows this function of series device. At 0.1s, utility voltage suddenly goes below to be 80 percents of previous voltage, as is shown in Fig.12. If series device keep running before voltage sag happen, utility voltage will keep almost const, as is shown in Fig.13.

| Items | Parameters |
|---|---|
| Utility fundamental voltage | 3-phase in positive sequence; line to line voltage: 10KV; Initial phase: 0 deg. |
| Utility 2nd harmonic voltage | 3-phase in negative sequence; line to line voltage: 250V; Initial phase: 0 deg. |
| Utility 3rd harmonic voltage | 3-phase in zero sequence; line to line voltage: 600V; Initial phase: 0 deg. |
| Utility 5th harmonic voltage | 3-phase in negative sequence; line to line voltage: 1500V; Initial phase: 0 deg. |
| Utility 7th harmonic voltage | 3-phase in positive sequence; line to line voltage: 1300V; Initial phase: 0 deg. |
| Impedance of transmission line | Resister: 0.04 ohm; Inductor : 1uH; |
| Low Pass filter | $L_1$: 4mH; $C_1$: 15uF |
| Transformer T2 | n=10 |
| Load | 3-phase series resister and capacitor Resister: 0.2 ohm; capacitor: 100uF |

Table 1. Parameters for series device

|  | 2nd (%) | 3rd (%) | 5th (%) | 7th (%) | THD(%) |
|---|---|---|---|---|---|
| Voltage before run | 3.07 | 7.35 | 12.24 | 9.79 | 17.58 |
| Voltage after run | 0.88 | 1.55 | 3.55 | 2.37 | 4.66 |
| current before run | 6.09 | 21.93 | 60.48 | 66.96 | 93.05 |
| current after run | 1.99 | 4.86 | 17.72 | 16.44 | 25.67 |

Table 2. Harmonics before and after series device run

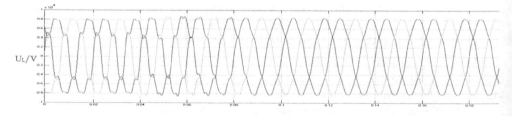

Fig. 7. Waveform of load voltage

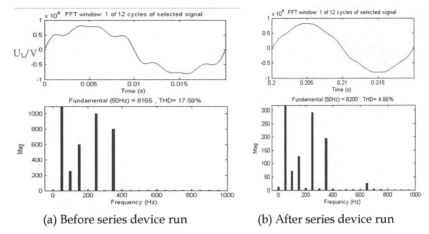

(a) Before series device run          (b) After series device run

Fig. 8. FFT analysis for load voltage

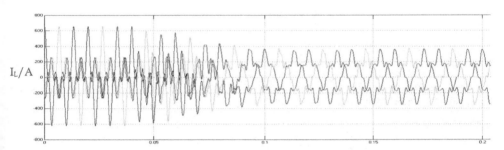

Fig. 9. Waveform of load current

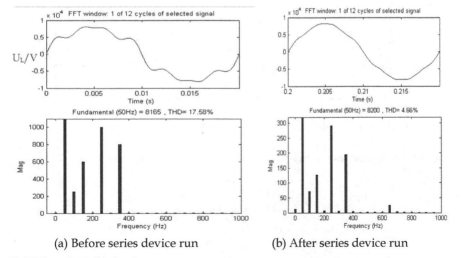

(a) Before series device run          (b) After series device run

Fig. 10. FFT analysis for load current

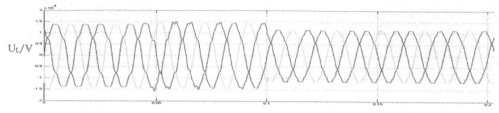

Fig. 11. Voltage sag at 0.1s

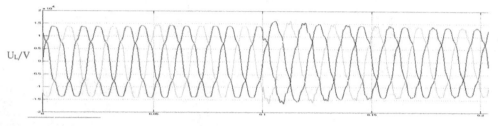

Fig. 12. Load voltage when series device run

### 3.2 Shunt device of high power UPQC

Fig.13 shows the single phase equivalent circuit of the shunt device of high power UPQC. The active part of the shunt device could be considered as an ideal controlled voltage source $U_{inv1}$, the Load harmonic source is equivalent to a current source $I_L$. The impedance of the output filter $L_0$ and $C_0$ are $Z_{L0}$ and $Z_{C0}$.

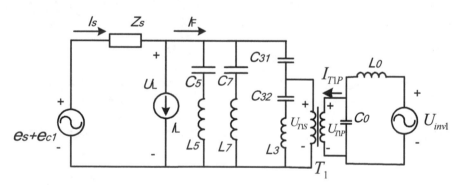

Fig. 13. The single phase equivalent circuit of the shunt device of UPQC

Suppose $\dfrac{U_{T1S}}{U_{T1P}} = n_1$ and transformer $T_1$ is a ideal transformer, we can learn

$$U_{inv1} = U_{T1P} + Z_{L0}(I_{T1P} + \frac{U_{T1P}}{Z_{C0}})$$

$$= U_{T1P}(1 + \frac{Z_{L0}}{Z_{C0}}) + Z_{L0}I_{T1P} \qquad (15)$$

And

$$I_{T1P} = n_1(I_F - \frac{U_L}{Z_{57}} - \frac{n_1 \cdot U_{T1P}}{Z_{332}}) \tag{16}$$

Where

$$Z_{57} = \frac{(Z_{L5} + Z_{C5})(Z_{L7} + Z_{C7})}{Z_{L5} + Z_{C5} + Z_{L7} + Z_{C7}}, \tag{17}$$

$$Z_{332} = Z_{L3} + Z_{C32} \tag{18}$$

Besides

$$U_{L1} - n_1 U_{T1P} = (I_F - \frac{U_L}{Z_{57}})Z_{C31} \tag{19}$$

So

$$U_{T1P} = \frac{Z_{57} + Z_{C31}}{n_1 Z_{57}} U_L - \frac{Z_{C31}}{n_1} I_F \tag{20}$$

From equation (16) and (20), we get

$$I_{T1P} = (1 + \frac{Z_{C31}}{Z_{332}})I_F - n_1(\frac{1}{Z_{57}} + n_1 \cdot \frac{Z_{57} + Z_{C31}}{n_1 Z_{57} Z_{332}})U_L \tag{21}$$

Where

$$Z_{57} = \frac{(Z_{C5} + Z_{L5})(Z_{C7} + Z_{L7})}{Z_{C5} + Z_{L5} + Z_{C7} + Z_{L7}} \tag{22}$$

$$Z_{332} = Z_{C32} + Z_{L3} \tag{23}$$

For completely compensating load harmonic current, $I_F$ is controlled to be the same as $I_L$, so

$$I_{T1P} = (1 + \frac{Z_{C31}}{Z_{332}})I_L - n_1(\frac{1}{Z_{57}} + n_1 \cdot \frac{Z_{57} + Z_{C31}}{n_1 Z_{57} Z_{332}})U_L \tag{24}$$

From equation (24), we can find control rule for shunt device of UPQC. If Inverter 1 is controlled to work as a current source, we can make it linear to load harmonic current and a fore-feed controller of load harmonic voltage is expected to add to the harmonic current controller. Control scheme for shunt device of high power UPQC is shown in Fig.14. To support DC linker voltage, shunt device should absorb enough energy from utility. Because it is easier for shunt device to absorb energy from utility, the DC linker voltage controller is placed in control scheme of shunt device. A PI conditioner is used here to adjust fundamental active current so as to keep DC-linker voltage const. ACR1 and ACR2 are the same as that of series device. Current out of active part is detected and form a close-loop controller. ACR3 is a hysteresis controller which makes Inverter 1 work as a current source. $U_L$ is also added to control scheme as a fore-feed controller.

Fig.15 shows the effect of this control scheme for shunt device of UPQC. The simulation parameters are shown in Tab.3. Suppose at 0.04s, passive part of shunt device is switched on and at 0.1s active part is started. Fig.15 shows waveform of utility current during shunt device is switched on. Fig.16 shows spectrums of utility current. Before shunt device switched on, THD of utility current is 28.53%. after passive part is switched on, it is cut down to be 18.25% and after active part is also switched on it is further cut down to be 11.97%.

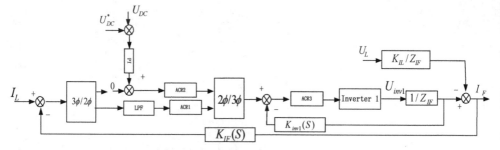

Fig. 14. Control scheme for shunt device of UPQC

| Items | Description |
|---|---|
| Power source | 3-phase; line to line voltage:10KV; |
| Impedance of transmission line | Resister: 0.04 ohm; Inductor : 1uH; |
| Load | Rectifier with series reactor and resister;<br>Reactor: 1mH; resister: 10 ohm; |
| Shunt device of UPQC | $3rd$ $L_3 = 15\,mH$ $C_{31} = 334\,\mu F$ $C_{32} = 669\,\mu F$<br>$5th$ $L_5 = 3.4mH$ $C_5 = 120\,\mu F$<br>$7th$ $L_7 = 1.5mH$ $C_7 = 140\,\mu F$<br>$T_1$ $n_1 = 10$<br>$LPF$ $L_0 = 4mH$ $C_0 = 15uF$ |

Table 3. Parameters for shunt device

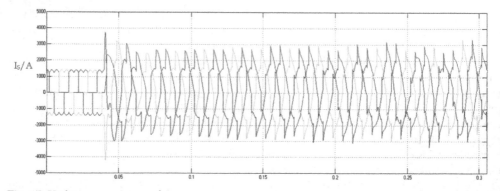

Fig. 15. Utility current waveform

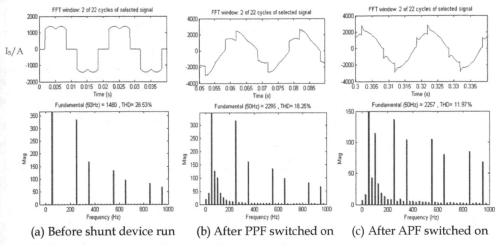

(a) Before shunt device run    (b) After PPF switched on    (c) After APF switched on

Fig. 16. Spectrums of utility current

### 3.3 Entire control of high power UPQC

High power UPQC is composed of series device and shunt device. Its control scheme combined control of series device and shunt device, as is shown in Fig.17. From above discussion, we know that load harmonic current is a bad disturb to series device controller because it influences load harmonic voltage. With shunt device, utility harmonic current is cut down and it does help to series device controller. On the other hand, load harmonic voltage is also a bad disturb to shunt device controller which will produce additional harmonic current and influence effect of shunt device. With series device, load harmonic voltage is cut down and it does help to shunt device controller. Cycling like this, effects of shunt device and series device are both improved. Tab.4 shows parameters for high power UPQC.

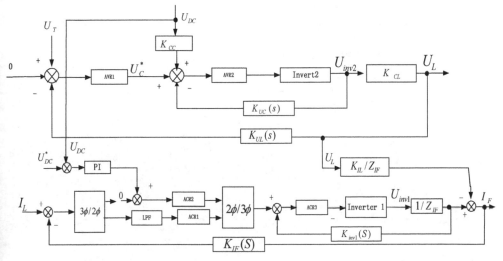

Fig. 17. Control scheme for high power UPQC

| Items | Description |
|---|---|
| Power source | 3-phase; line to line voltage:10KV; 2nd, 3rd, 5th, 7th harmonic voltage listed in Tab.1 |
| Impedance of transmission line | Resister: 0.04 ohm; Inductor : 1uH; |
| Load | Rectifier load in Tab.3 paralleled with 3-phase series resister and capacitor listed in Tab.1 |
| Shunt device | Same as Tab.3 |
| Series device | Same as Tab.1 |

Table 4. Parameters for high power UPQC.

Suppose at 0.04s, series device is switched on, at 0.1s passive part of shunt device is switched on and finally at 0.16s active part of shunt device is also switched on. Fig.18 shows the utility current waveform and Fig.19 shows its spectrums. Fig.20 shows the utility voltage waveform and Fig.21 shows its spectrums. The harmonics during switching on the whole UPQC are shown in Tab.5. We can see that power quality is improved step by step.

| THD(%) | Before UPQC run | Series device only | Switch on passive part | Switch on active part |
|---|---|---|---|---|
| Utility voltage | 17.7 | 8.26 | 4.78 | 4.77 |
| Utility current | 40.36 | 31.10 | 11.97 | 8.90 |

Table 5. THD comparison during switching on UPQC

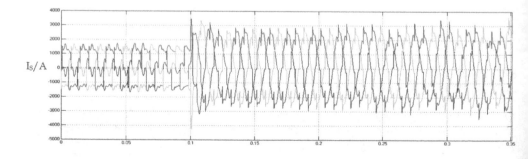

Fig. 18. Utility current waveform

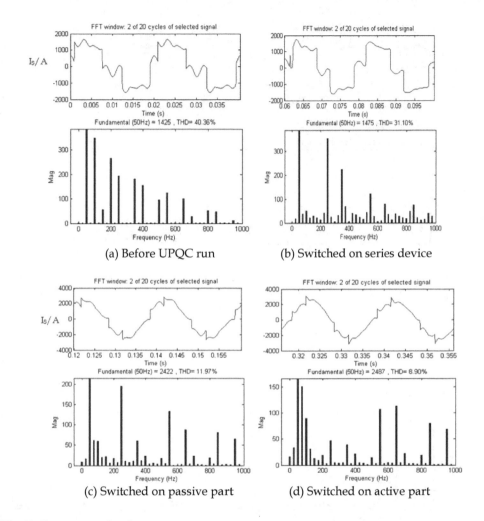

Fig. 19. Spectrums of utility current

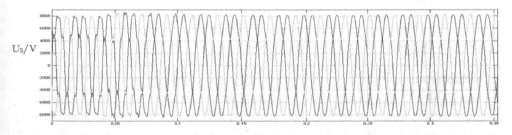

Fig. 20. Utility voltage waveform

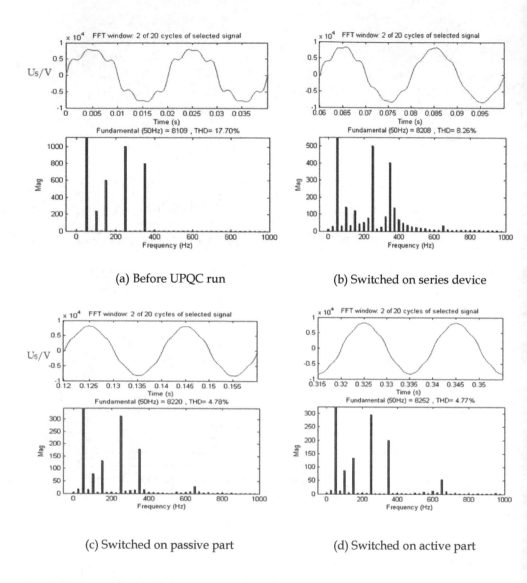

(a) Before UPQC run                    (b) Switched on series device

(c) Switched on passive part            (d) Switched on active part

Fig. 21. Spectrums of utility voltage

## 4. Conclusions

To eliminate harmonics in power system, series APF and shunt APF are adopted. Series APF mainly eliminate harmonic voltage and avoid voltage sag or swell so as to protect critical load. It also helps to eliminate harmonic current if power source voltage is distorted. Shunt APF is to eliminate harmonic current avoiding it flowing through transmission line. UPQC combined series APF and shunt APF can not only eliminate harmonic current but also guarantee a good supply voltage.

In some applications, the equipment needs to compensate high power reactive power produced by load. In this case, An UPQC with current-injection shunt APF is expected to be installed. This chapter discussed the principle of UPQC, including that of its shunt device and series device, and mainly discussed a scheme and control of UPQC with current-injection shunt APF which can protect load from almost all supply problems of voltage quality and eliminate harmonic current transferred to power grid.

In high power UPQC, load harmonic current is a bad disturb to series device controller. Shunt device cuts down utility harmonic current and does help to series device controller. On the other hand, load harmonic voltage is also a bad disturb to shunt device controller and series device does much help to cut it down. With the combined action of series device and shunt device, high power can eliminate evidently load harmonic current and harmonic voltage and improve power quality efficiently.

## 5. References

Terciyanli, A., Ermis, M.& Cadirci, I. (2011). A Selective Harmonic Amplification Method for Reduction of kVA Rating of Current Source Converters in Shunt Active Power Filters, Power Delivery, Vol.6., No.1, pp.65-78, ISSN: 0885-8977

Wen, H., Teng, Z., Wang, Y. & Zeng, B.(2010). Accurate Algorithm for Harmonic Analysis Based on Minimize Sidelobe Window, Measuring Technology and Mechatronics Automation , Vol.1., No.13-14, pp.386-389, ISBN: 978-1-4244-5001-5

Ahmed, K.H., Hamad, M.S., Finney, S.J., & Williams, B.W.(2010). DC-side shunt active power filter for line commutated rectifiers to mitigate the output voltage harmonics, Proceeding of Energy Conversion Congress and Exposition (ECCE), 2010 IEEE, pp.151-157, ISBN: 978-1-4244-5286-6, Atlanta, GA, USA, Sept.12-16, 2010

Wu, L.H., Zhuo, F., Zhang P.B., Li, H.Y., Wang, Z.A.(2007). Study on the Influence of Supply-Voltage Fluctuation on Shunt Active Power Filter, Power Delivery, Vol.22, No.3, pp.1743-1749, ISSN: 0885-8977

Yang, H.Y., Ren, S.Y.(2008), A Practical Series-Shunt Hybrid Active Power Filter Based on Fundamental Magnetic Potential Self-Balance, Power Delivery, Vol.23, No.4, pp.2089-2192, ISSN:0885-8977

Kim, Y.S., Kim, J.S., Ko, S.H.(2004). Three-phase three-wire series active power filter, which compensates for harmonics and reactive power, Electric Power Applications, Vol.153, No.3, pp.276-282, ISSN: 1350-2352

Khadkikar, V., Chandra, A., Barry, A.O., Nguyen, T.D.(2005). Steady state power flow analysis of unified power quality conditioner (UPQC), ICIECA 2005. Proceeding of International Conference, pp.6-12, ISBN: 0-7803-9419-4, Quito, May 10-14, 2005

Brenna, M., Faranda, R., Tironi, E.(2009). A New Proposal for Power Quality and Custom Power Improvement: OPEN UPQC, Power Delivery, Vol.24, No.4, pp.2107-2116, ISSN:0885-8977

Zhou, L.H., Fu, Q., Liu, C.S.(2009). Modeling and Control Analysis of a Hybrid Unified Power Quality Conditioner, Proceeding of 2009. Asia-Pacific Power and Energy Engineering Conference, pp.1-5, ISBN: 978-1-4244-2486-3 , Wuhan, March 27-31, 2009

# 7

# Stochastic Analysis of the Effect of Using Harmonic Generators in Power Systems

Mohsen Abbas Pour Seyyedi and Amir Hossein Jahanikia
*Mefragh Company*
*Iran*

## 1. Introduction

Switch mode electronic devices including Compact Fluorescent Lamp (CFL) and personal computers introduce capacitive power factor and current harmonics to the power system. Since middle 80's and with the expanding use of nonlinear switch mode electronic loads, concerns arose about their effect on the power systems. In many IEEE documents, it is recommended to study the effect of electronic loads. Switch mode devices have a capacitive power factor between 55 and 93 percent (Allexperts), which can cause the increase of reactive power and power loss. The power loss in an office building wirings due to the current harmonics may be more than twice that of the linear load equipment (Key et al., 1996). Capacity of the transformers may be reduced more than 50 per cent in the presence of harmonic components (Schneider, 2009).

CFL is a more efficient and durable replacement of the traditional incandescent lamp. Replacing traditional light bulbs by CFLs has several advantages including energy saving, increase in the capacity of plants and distribution transformers, peak shaving, less carbon emission and customer costs. On average, 20 percent of the total use of electricity is consumed in lighting (Michalik et al., 1997), (Tavanir). However, the increase in the number of electronic devices especially the CFLs in power systems must be carefully planned. Replacing the incandescent light bulbs with CFLs means replacing the system's major Ohmic load with a capacitive load of high frequency harmonic components. In areas where lighting is a major use of electricity, e.g. places where natural gas or other fossil fuels are used for heating purposes, unplanned replacing of incandescent lamps with CFLs can introduce unexpected negative effects on the system. Also, in areas with a considerable number of other switch mode devices e.g. commercial areas with many office buildings it is important to plan the number of CFLs carefully. Most of the present studies on the effect of switch mode devices are based on tentative experiments and power factor measuring before and after using the devices in the power system (Gonos et al., 1999), and proposing a model for the network has been less discovered.

In order for studying such effects, it is better to classify the system equipment to the substation equipment and consumer side equipment. Dramatic changes in power quality indicators of the distribution systems may cause disorders or even damages in the consumer equipments. Such disorders are especially important for sensitive appliances such as medical and hospital devices.

In this chapter we review our novel approach for studying the effect of switch mode devices and present a novel stochastic modelling approach for analysing the behaviour of the power system in the presence of switch mode devices. We also study the major KPI of the power system and study how these KPI will be affected by adding the current harmonics. Section 2 presents how we obtain an accurate model for CFL based on circuit simulation. This section also defines a general circuit model for the harmonic generating devices. Section 3 presents our novel approach for stochastic modelling of the power system behaviour. In section 4 we summarize the major power system KPI on both substation and consumer sides. We also discuss how the switch mode devices may affect the devices on each side. Section 5 presents our approach for simulating the power system behaviour. Conclusion and discussion are presented in section 6.

## 2. Modelling of switch mode devices

This section studies the general specifications of switch mode devices. We simulate a CFL ballast circuit in SPICE software. We also present the device model for a personal computer. Based on these models, we develop a general circuit model to simulate the behaviour of all switch mode capacitive devices. Without circuit simulation, it is not possible to provide an accurate model representation in the power system. In contrast with the models that are based on measuring and estimating the device characteristic, this approach gives much more accurate results. The accuracy of this approach can be chosen at the desired level.

### 2.1 Simulation of CFL ballast circuit in SPICE
The common 220V power system voltage is not enough to start the fluorescent lamps. Therefore, CFLs include a ballast circuit for providing the starting high voltage. In traditional fluorescent lamps, inductive ballasts are widely utilized. However, electronic ballasts which are used in CFLs have much better quality (Aiello et al., 2008). Electronic ballasts are composed of a rectifier and a DC-AC converter. Fig. 1 shows the general block diagram of a ballast circuit.

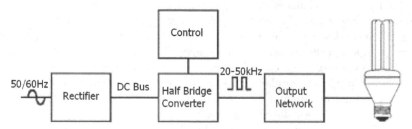

Fig. 1. Block diagram of a CFL ballast circuit. Figure courtesy of (Sasaki, 1994).

Several circuits are simulated in SPICE software for this project. Fig. 2 shows one sample CFL ballast circuit model in SPICE. This circuit is similar to that of (Sasaki, 1994) with slight changes. The input full wave rectifier and the large input capacitor make the current have narrow high peaks at short intervals and almost zero value elsewhere. Fig. 3 shows the output voltage and current of the circuit in Fig. 2.

Frequency analysis shows that the CFL current is made up of odd harmonic components of the main frequency (50 or 60 Hz). The CFL is modelled by a number of current sources with the proper harmonic values. Equation 1 shows the mathematical model for a CFL when the voltage is assumed to be a cosine function.

$$v_{CFL} = V \cos(2\pi f t)$$

$$i_{CFL} = \sum_{n=0}^{\infty} I_{2n+1} \cos[2\pi(2n+1)ft + \Phi_{2n+1}] \approx \sum_{n=0}^{4} I_{2n+1} \cos[2\pi(2n+1)ft + \Phi_{2n+1}]$$

(1)

The more the number of harmonics is, the more accurate the model will be. In this study we use the first five odd harmonics (1, 3, 5, 7, and 9). A schematic of the model is shown in Fig. 4. The power factor of this circuit is 93%. In order for having a flexible model for different market suppliers, the power factor is chosen flexible in the simulation experiments.

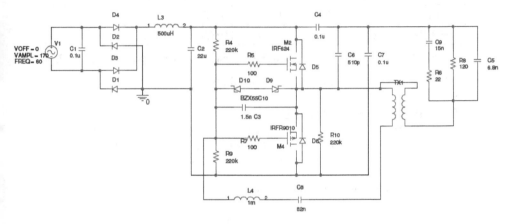

Fig. 2. Simulation of a sample ballast circuit in SPICE.

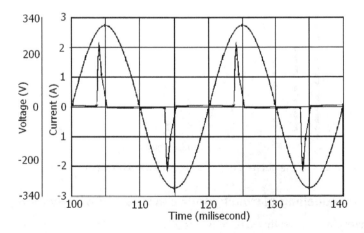

Fig. 3. Sinusoidal voltage and resulting current waveshape for a sample CFL ballast circuit.

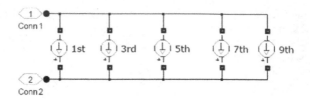

Fig. 4. Circuit model of a switch mode device.

The values of the current and phase in equation 1 are summarized in Table 1 for the circuit in Fig. 2.

| Current Harmonic | First | Third | Fifth | Seventh | Ninth |
|---|---|---|---|---|---|
| Peak $I_{2n+1}$ (A) | 0.2 | 0.182 | 0.162 | 0.138 | 0.112 |
| Phase $\Phi_{2n+1}$ (Rad) | 0.260 | 3.499 | 0.609 | 4.000 | 0.799 |

Table 1. Peak value and phase of the current harmonics for the sample CFL of Fig. 2.

We name the overall current phase lag as central phase lag $\Phi_c$.

## 2.2 Circuit model for other electronic devices
Personal computers and other electronic equipment such as printers, etc. generate current harmonics in the power system too, because they all include a rectifier. The harmonic components of personal computers are calculated and provided in the literature (Key et al., 1996). Fig. 5 shows the relative value of these components. Therefore, we can use a similar model to that of Fig. 4 for modelling such electronic devices.

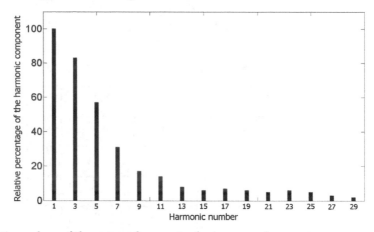

Fig. 5. Relative values of the current harmonics for a personal computer.

## 3. Stochastic modelling of switch mode devices in power system
Phase of a harmonic generating device is not a constant value. But it is a random variable that varies in a specific range that can be provided by the manufacturer. Therefore, the model in equation 1 will be modified to that of equation 2.

$$i_m = \sum_{n=0}^{4} I_{2n+1} \cos\left[2\pi(2n+1)ft + \Phi_{2n+1} + (2n+1)\times\Delta\Phi_m\right] \tag{2}$$

In this model, $m$ is the device number in the network. In practice, the total number of devices $M$ is a large number. For each device we have a phase shift $\Delta\Phi_m$ from the central phase. Distribution of $\Delta\Phi_m$ can be assumed to be uniform or Gaussian depending on the manufacturer's datasheet. In other words for the uniform distribution:

$$\Phi_c - \Phi_{max} < \Delta\Phi_m < \Phi_c + \Phi_{max}, \qquad P(\Delta\Phi_m = \phi) = \frac{1}{2\Phi_{max}} \tag{3}$$

In the above equation, $\Phi_{max}$ is the maximum phase shift from the theoretical phase lag $\Phi_c$. If the probability distribution is Gaussian, $\Delta\Phi_m$ is obtained from equation 4:

$$P(\Delta\Phi_m = \phi) = \frac{1}{\sqrt{2\pi\sigma^2}} e^{(\phi-\bar{\phi})^2/2\sigma^2} \tag{4}$$

Where $\varphi$ and $\sigma$ are the mean and variance of the phase lag. The current value for a large number $M$ of CFLs with the above specifications is equal to $i$ in equation 3:

$$i = \sum_{m=0}^{M} i_m = \sum_{m=0}^{M}\sum_{n=0}^{4} I_{2n+1} \cos\left[2\pi(2n+1)ft + \Phi_{2n+1} + (2n+1)\times\Delta\Phi_m\right] \tag{5}$$

Finding the probability density function, expectation and variance of current in the above equation is complicated (if possible!). Instead, we rely on numerical simulation to find the *pdf* of power system current. In a sample experiment, our power system is composed of a thousand CFLs. The average phase lag of these CFLs is fifteen degrees and has a uniform distribution of range *15±10* degrees. We use a Parzen window (Duda et al., 2007) approach to find the *pdf* of the peak output current. Fig. 6 shows the result. The mean value and standard deviation of the current in this experiment are 611.5 and 2.45. We may notice that if the variance in the phase shift is not considered, meaning that we do not assume the CFL phase to be a random variable, the peak current will be equal to 701 Amperes. When we dimension the network, this means at least 13% more capacity. Different results can be obtained for different values of phase range for the electronic devices.

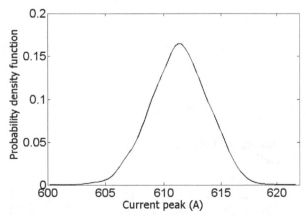

Fig. 6. Relative values of the current harmonics for a personal computer.

## 4. Effect of switch mode devices in power system

The main devices in any power systems include thermal loads, electronic devices and inductive loads. Thermal loads, including the traditional lighting, can be modelled as a simple resistance. In regions where a considerable part of heating is based on electricity, heating constitutes a large share of the thermal load. Electronic devices include television and personal computer. The most common inductive appliances are cooling devices, which are air conditioning systems and refrigerators. Different values of power from 150W to 2000W are used for induction motors, depending on the application type and geographical region of use.

In this section we take a closer look to the different equipment and the effects of the harmonics on individual devices on both distribution and consumer sides. This section mainly focuses on theoretical study of the effects of the harmonics on the key performance indicators rather than practical measurements. Fig. 2 shows the fundamentals of the analysis in the rest of this section.

Fig. 7. Relation between using CFL (as a switch mode device) and the power system equipment.

### 4.1 General power system KPI
The most important and most useful KPI of the power system include:

### 4.1.1 Total Harmonic Distortion (THD)
THD is defined as in equation 6:

$$THD = \frac{\sqrt{\sum_{n=2}^{\infty} V_n^2}}{V_1}$$

(6)

This factor may be calculated for voltage, current or power. Current THD is normally much higher than the voltage THD for electronic devices and is more than 100%.

### 4.1.2 Crest Factor
Crest Factor is the indicator that shows the ratio of the peak value of the wave shape to its average value in a time interval (rms)[1], as shown in equation 7.

$$C = \frac{|x|_{peak}}{x_{rms}} \tag{7}$$

This factor is especially important for the analysis of the effect on protective and control equipment such as relays.

### 4.1.3 Power Factor
Power Factor is caused by the input rectifying capacitor of the ballast circuit. The capacitive characteristic of CFL is an advantage in systems with a high inductive load.

### 4.2 Power substation equipment
This section studies the effect electronic devices on the substation equipments, which provide the low voltage power to the end users and can be classified to the following categories:

### 4.2.1 Transformers
Transformers are used in the distribution system in order to change the levels of voltage and current in the low voltage scales. These may also include the power and instrumentation transformers and Auto-Boosters. In transformers, both the core and the wires are sensitive to the change of the power KPI. The harmonics have the following different effects on transformers.

*Transformer loss*, which is obtained as in equation 8:

$$P_T = P_N + P_{LL} \tag{8}$$

In equation 8, $P_N$ is the "no-load" loss and $P_{LL}$ is the "full-load" loss. The no-load loss depends on the voltage and core material. The full load loss is defined as in equation 9:

$$P_{LL} = P_{DC} + P_{EC} + P_{OSL} \tag{9}$$

In equation 9, $P_{DC}$ is the DC resistance loss, $P_{EC}$ is the eddy current loss and $P_{OSL}$ is the stray loss. Eddy current, which is proportional to the square of frequency, is caused by skin effect and proximity effect. Therefore, the current harmonics increase the eddy current loss dramatically. This increase results in the increase in temperature and hence reducing the transformer lifetime (Ashok).

*Lifetime* of a transformer depends on the functioning situations such as loading percentage and functioning temperature. Current harmonic components can increase the RMS value of

---

[1] Root Mean Square

the current and consequently the resistive power loss. The heat also remains in the surrounding air and affects the transformer lifetime. In the delta-wye connections in the transformers, the harmonic components of current start rotating in the wye side and cause heat generation and reduce transformer capacity. This occurs for the 3rd, 9th and 15th harmonic components. The current harmonics also cause saturation of the transformer.

In order to control the effect of current harmonics on transformers and electromotors, it is recommended not to have a current harmonic component more than 5% of the transformer's nominal current in the ANSI/IEEE C57.12.00-2000 and IEC60076 standards (Sadati et al., 2008). The K-Factor, which is defined in equation 10, identifies the relation between transformer design and increase in the electronic devices.

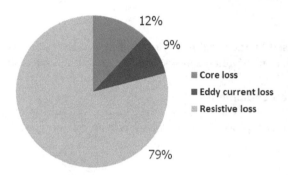

Fig. 8. Comparing different losses in the transformer core.

$$K - Factor = \sum_{h=1}^{\infty} I_h (ph)^2 h^2 \qquad (10)$$

In equation 10, $h$ is the harmonic order and $I_h$ is its current component in per unit. The definition of K-Factor may be different in different standards. For example, in BS7821 standard it is defined as in equation 11.

$$K - Factor = \left[ 1 + \frac{e}{1+e} \left( \frac{I_1}{I} \right)^2 \sum_{n=2} \left( n^q \left( \frac{I_n}{I_1} \right)^2 \right) \right]^{0.5} \qquad (11)$$

### 4.2.2 Transmission systems

Transmission systems include power cables and wires. Transmitters are the conductors which are used in power system. Because of the skin effect (Hightech), the resistance of conductors increases with the increase of frequency. Because of the small high frequency components of the electronic devices, using these devices has a small effect on the conductors. Simulations show that the current amplitude for a sample CFL in frequencies between 1000 and 1500 Hertz (harmonics 20th to 30th) is less than 1% of the nominal value. Therefore, capacity reduction of cables and wires will be between 1% and 6% of conductor capacity in the worst case (100% replacement) (Vapopoulos, 1964).

### 4.2.3 Electronic devices:
Electronic devices including the automation or control devices and the equipments that are used for stabilizing the power systems such as PLC transmitters and RTU equipment. As the electronic devices are used for transmitting and receiving data, high frequency current can cause disorders in their functioning, such as increased noise level in communication systems. Such devices are usually installed close to power system equipment such as transformers.

### 4.2.4 Control and protection systems
Control and protection system, such as fuses, relays and circuit breakers, which control or guard the power systems. Current harmonics in the system may cause pre-heating in the fuse and problems in its function. Fuses may also be affected by the skin effect and the resulting heat may cause their malfunctioning. In circuit breakers, which work based on $di/dt$, current harmonics may cause unexpected faults. Here, the peak factor is important as well. The solenoids may also be damaged because of the harmonics. Delay in solenoid's functioning may cause sparks and damage. Vacuum circuit breakers are less sensitive to the harmonics. For relays, changes in the zero point may case improper functioning. These effects must be identified via practical experiments. Some manufacturers have presented "harmonic adaptive" models (Ashok).

### 4.3 Equipment on the consumer side
The consumer side equipments are classified into three different categories (Gowan, 2006):

### 4.3.1 Electric machines
Electric machines which are inductive devices and as confirmed by simulation using them together with electronic devices can improve the power quality indicators. The $n_{th}$ current component of an electric machine is obtained as in equation 12:

$$I_n = \frac{V_n}{n \cdot f \cdot L_n} \tag{12}$$

In equation 12, $V_n$ is the $n_{th}$ voltage harmonic and $L_n$ is the inductance in $n_{th}$ harmonic (Markiewicz et al., 2004). $L_n$ increases because of the skin effect. Power loss of electric motors is obtained from equation 13:

$$P_h = \Sigma P_n = kP_r \Sigma \left( \frac{V_n^2}{n^{1.5} \cdot V_1^2} \right) \tag{13}$$

where $k = \frac{T_s}{T_r} \cdot \frac{E}{(1-S_r)(1-E_r)}$ . $T_s$ is the torque, $E_r$ is efficiency and $S_r$ is the slippage (Schneider 2010).

The heat generated by the harmonics may cause decreasing the motor lifetime. In addition, the eddy currents can generate heat in the motors, similar to the transformers. The heat generation is almost the same in synchronous and asynchronous motors. In the case when all incandescent lamps are replaced by CFLs, the performance of induction motors will be between 5 and 15 percent (Vapopoulos, 1964). For the asynchronous motors, IEC60892 standard is defined as in equation 14:

$$HVF = \sqrt{\sum_{h=2}^{13} \frac{U_h}{h^2}} \leq 0.02 \qquad (14)$$

$U_h$ is the harmonic voltage of order $h$.

### 4.3.2 Measurement devices

Measurement devices: such as electricity meters, current transformers, voltage transformers and electronic instrumentation. The measurement transformer error is calculated as:

$$K_f = CVK_p \qquad (15)$$

where $C$ is the ratio of real ratio to actual ratio (for current transformers), $V$ is the ratio of real ratio to actual ratio (for voltage transformers) and $K_p$ is angle correction coefficient (Hightech). The same as in power station transformers, high frequency current harmonics result in early saturation of transformer core and error in measurement. Electronic devices which work based on the zero crossing may also be affected by wave shape distortion. Errors may occur while calculating voltage and current *rms* values. In the inductive electricity meters (Temple, 1998) the error may be more. Changes in power factor and the THD affect the operation of these equipments.

### 4.3.3 Communication devices

Most communication systems are equipped with filters to reduce the noise generated by high frequency components. As the harmonics above 1500 Hz are negligible in electronic devices, they will practically not affect the communication systems. Similarly they will not affect electronic devices on the consumer side. Some malfunction of printers is reported (Abbaspour et al., 2009). Multimedia devices and televisions may be affected when used in the presence of harmonics. Distortion may be introduced by either the harmonics or the electronic ballast frequencies. All such devices are equipped with electromagnetic filters (EMC) which remove such effects. Personal Computers are sensitive to a voltage distortion of more than 5%. However, the use of harmonic generating devices does not introduce more than 0.5% voltage THD.

As the CFL ballast circuits work in 40 KHz frequency, they may affect hospital devices. Some distortions are reported in the functioning of CT devices (Abyaneh, 2004). Also because of the type of the CFL light, it is not recommended in operation rooms. The most important quality parameters which are considered here are the total power use, total harmonic distortion and transformer eddy current loss and hysteresis effect. Different measurements can explain the effect of harmonics. Total harmonic distortion (THD) is defined in equation 1 (Chapman).

## 5. Simulating of power system

In this section, we develop individual units of consumers developed, which are a certain combination of electric device models. These units are called with terms "home" and "office". These models are obtained using statistical data about different residential regions in Iran. In the simplest case, a home is a combination of four lighting loads and one induction motor. Fig. 9 shows this consumer unit simulated in Simulink (Mathworks).

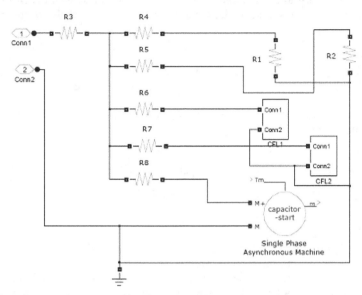

Fig. 9. "Home" unit with two CFLs, two lamps and induction motor.

Fig. 10 shows a sample commercial (office) consumer unit.

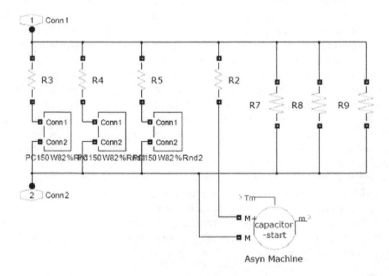

Fig. 10. "Office room" unit with four incandescent lamps, three PCs and a single phase asynchronous motor.

Eddy current loss of transformer core depends on the squares of both current and frequency (Bird). Therefore, the core loss for different experiments is compared to each other using equation 16. $f_{h1}$ and $f_{h2}$ correspond to the frequency components of the different cases in comparison.

$$\text{Relative core loss} = \frac{\sum_{h=1}^{N} f_{h2}^2 I_{h2}^2}{\sum_{h=1}^{N} f_{h1}^2 I_{h1}^2} \tag{16}$$

A similar approach is used for the hysteresis effect. However, hysteresis effect comparison is obtained using equation 17 (Bird).

$$\text{Relative hysteresis} = \frac{\sum_{h=1}^{N} f_{h2} I_{h2}^{1.6}}{\sum_{h=1}^{N} f_{h1} I_{h1}^{1.6}} \tag{17}$$

## 6. Conclusion

A new approach is presented for the analysis of switch mode devices and their effect on distribution network. In this approach, model of electronic devices are derived from detailed circuit simulation and the results are transformed to a circuit model which is a combination of passive current circuits. The model contains the odd harmonics of the power system frequency. Based on this model, we developed a probabilistic model for the electronic device and used numerical calculation to estimate the probability distribution of the power system current as a whole. This model shows that adding electronic devices to the power system must be carefully planned so that the power system KPI do not exceed the recommended levels. This stochastic simulation is a strong tool to evaluate the situation of the power system and its capacity for growth. In case of mass use of electronic devices, additional capacity must be provided both for transformers and distribution lines, in order to avoid harms in the network via harmonic mitigation.

We also presented a list of the power system KPI and the devices that can be affected by the additional current harmonics. We classified the equipments into the feeding and consuming sides and studied the effects on each side separately. Mass usage of switch mode devices in different power systems must be planned carefully in order to avoid any unexpected negative effects on the other equipments in the system. The most vulnerable equipments of the power system are the ones which contain solenoids, such as transformers and measurement equipments, especially on the feeding side. Therefore, extra care and calculations are required in different power systems for a safe use of electronic devices. Electronic devices are affected by the harmonic distortion and may need to be either equipped with protective filters or be replaced by more advanced ones. Based on the power system KPI and the specification of the distribution and consumption devices we can use the stochastic analysis approach for dimensioning an optimal power system with respect to equipment costs and safety margins.

## 7. Acknowledgment

The authors of this paper wish to acknowledge the support of Mefragh Company, Tehran Iran, especially Mr. Hashem Khoee, president of the company. We also wish to thank Iran

Energy efficiency Organization (IEEO - SABA) and especially Mr. Sajjadi for their continual support.

# 8. References

Key, T. S.; Jih-Sheng, L., (1996). Costs and benefits of harmonic current reduction for switch mode power supplies in a commercial office building, IEEE Tran. on Industry Applications, vol. 32, Issue 5, Sep/Oct 1996, pp. 1017–1025.

Schneider Electric. Electrical Installation Guide, 2009.

Michalik, G.; Khan, M. E.; Bonwick, W. J.; Mielczarski, W. (1997). Structural modeling of energy demand in the residential sectors. Elsevier, vol. 38, Number 6, November 1997, pp. 453–453.

http://amar.tavanir.org.ir/

Gonos, I. F.; Kostic, M. B. ; Topalis, F. V. (1999). Harmonic distortion in electric power systems introduced by compactfluorescent lamps. International Conference on Electric Power Engineering, Budapest, 1999, pp. 295–300

Aiello, N.; Messina, S. (2008). VIPower: Self-Oscillating Converter Using VK05CFL for Compact Fluorescent Lamps.

Sasaki, R. I. (1994). The Impact of Electronic Ballast Compact Fluorescent Lighting on Power Distribution Systems. Purdue University School of Electrical Engineering, 1994.

Duda, R. O.; Hart, P. E.; Stork, D. G. (2007) Pattern classification. Wiley, ISBN 978-0-471-05669-0

Sadati, S. B. ; Tahani, A. ; Darvishi, B. ; Dargahi, M. ; Yousefi, H. (2008). Comparison of Distribution Transformer Losses and Capacity under Linear and Harmonic Loads. in 2008 2nd IEEE International Conference on Power and Energy (PECon 08), Johor Baharu, Malaysia.

Ashok, S. Effect of power system harmonics on power system equipment. Nalanda Library Lectures.

Hightech Aus Co. Power factor correction and harmonics.

Vapopoulos, N. Th. (1964). Studies on harmonic analysis," Proc. Camb. Phil. Soc., 1964, pp. No 60, vol 465.

Gowan, C., Power quality and harmonics," Cornell University, 2006.

Markiewicz, H. Klajn ; A. (2004). Power quality application guide. Wroclaw University of Technology, July 2004.

Schneider-Electric Publication (2010). The Electrical Installation Guide (EIG), ISBN: 978.2.9531643.3.6, version 2010.

Temple, D. (1998). Facilities Instructions, standards & techniques. United States Department of the Interior Bureau of Reclamation, Power O&M Bulletin No. 13, vol. 2-2, 1998.

Abbaspour, M. ; Jahanikia, A. H. (2009). Power Quality Consideration in the Widespread Use of Compact Fluorescent Lamps. in Proc. 2009 10th Int. Conf. Electrical Power Quality and Utilisation, Lodz, Poland, pp. 1-6.

Abyaneh, H. A. (2004). Analysis of harmonic distortion in power distribution systems. presented at the 9th Int. Conf. Power Distribution Systems, Tehran, Iran, 2004 (in Persian).

Chapman, D. Rating of Transformer supplying harmonic loads. Leonardo Energy.

Mathworks. www.mathworks.com

Bird, J. O. Electrical circuit theory and technology. Newnes, May.

# 8

# Characterization of Harmonic Resonances in the Presence of the Steinmetz Circuit in Power Systems

Luis Sainz[1], Eduardo Caro[2] and Sara Riera[1]
*[1]Department of Electrical Engineering, ETSEIB-UPC,*
*[2]Department of Electrical Engineering, GSEE-UCLM,*
*Spain*

## 1. Introduction

An electric power system is expected to operate under balanced three-phase conditions; however, single-phase loads such as traction systems can be connected, leading to unbalanced line currents. These systems are single-phase, non-linear, time-varying loads closely connected to the utility power supply system. Among problems associated with them, special consideration must be given to the presence of unbalanced and distorted currents (Barnes & Wong, 1991; Capasso, 1998; Hill, 1994; Howroyd, 1989; Marczewski, 1999; Qingzhu et al., 2010a, 2010b). These operating conditions damage power quality, producing undesirable effects on networks and affecting the correct electric system operation (Arendse & Atkinson-Hope, 2010; Chen, 1994; Chen & Kuo, 1995; Chindris et. al., 2002; Lee & Wu, 1993; Mayer & Kropik, 2005). The unbalanced currents cause unequal voltage drops in distribution lines, resulting in load bus voltage asymmetries and unbalances (Chen, 1994; Qingzhu et al., 2010a, 2010b). For this reason, several methods have been developed to reduce unbalance in traction systems and avoid voltage asymmetries, for example feeding railroad substations at different phases alternatively, and connecting special transformers (e.g. Scott connection), Static Var Compensators (SVCs) or external balancing equipment (ABB Power Transmission, n.d.; Chen, 1994; Chen & Kuo, 1995; Hill, 1994; Lee & Wu, 1993; Qingzhu et al., 2010a, 2010b). The last method, which is incidentally not the most common, consists of suitably connecting reactances (usually an inductor and a capacitor in delta configuration) with the single-phase load representing the railroad substation (Barnes & Wong, 1991; Qingzhu et al., 2010a, 2010b). This method is also used with industrial high-power single-phase loads and electrothermal appliances (Chicco et al., 2009; Chindris et. al., 2002; Mayer & Kropik, 2005).

This delta-connected set, more commonly known as Steinmetz circuit, (Barnes & Wong, 1991; Jordi et al., 2002; Mayer & Kropik, 2005), allows the network to be loaded with symmetrical currents. Several studies on Steinmetz circuit design under sinusoidal balanced or unbalanced conditions aim to determine the reactance values to symmetrize the currents consumed by the single-phase load. Some works propose analytical expressions and optimization techniques for Steinmetz circuit characterization, (Arendse & Atkinson-Hope, 2010; Jordi et. al, 2002; Mayer & Kropik, 2005; Qingzhu et al., 2010a, 2010b; Sainz & Riera, submitted for publication). In general, the values of the symmetrizing elements should vary in order to compensate for the usual single-phase load fluctuations. Unfortunately, the typical inductances and capacitors have fixed values. However, this can be solved by the introduction of thyristor-controlled reactive elements due to the development of power electronics in the last few years and the use of step variable capacitor banks, (Barnes & Wong, 1991; Chindris et al., 2002).

Steinmetz circuit design must consider circuit performance and behavior under non-sinusoidal conditions because of the growing presence of non-linear devices in electric power systems in the last few decades, (Arendse & Atkinson-Hope, 2010; Chicco et al., 2009; Czarnecki, 1989, 1992). Harmonic currents injected by non-linear devices can cause voltage distortions, which may damage power quality. In this sense, the effects of harmonics on power systems and their acceptable limits are well known [IEC power quality standards, (IEC 6100-3-6, 2008); Task force on Harmonic Modeling and Simulation, 1996, 2002]. In the above conditions, the parallel and series resonance occurring between the Steinmetz circuit capacitor and the system inductors must be located to prevent harmonic problems when the Steinmetz circuit is connected. The parallel resonance occurring between the Steinmetz circuit capacitor and the supply system inductors is widely studied in (Caro et al., 2006; Sainz et al., 2004, 2005, 2007). This resonance can increase harmonic voltage distortion in the presence of non-linear loads injecting harmonic currents into the system. The problem is pointed out in (Sainz et al., 2004). In (Caro et al., 2006; Sainz et al., 2005), it is numerically and analytically characterized, respectively. In (Sainz et al., 2005), several curves are fitted numerically from the power system harmonic impedances to predict the resonance at the fifth, seventh and eleventh harmonics only. In (Caro et al., 2006), the resonance is analytically located from the theoretical study of the power system harmonic impedances. Finally, the analytical expressions in (Caro et al., 2006) to predict the parallel resonance frequency are expanded in (Sainz et al., 2007) to consider the influence of the Steinmetz circuit capacitor loss with respect to its design value. The series resonance "observed" from the supply system is also studied and located in (Sainz et al., 2009a, 2009b, in press). This resonance can affect power quality in the presence of a harmonic-polluted power supply system because the consumed harmonic currents due to background voltage distortion can be magnified. It is numerically and analytically studied in (Sainz et al., 2009a, 2009b), respectively. In (Sainz et al., 2009a), graphs to locate the series resonance frequency and the admittance magnitude values at the resonance point are numerically obtained from the power system harmonic admittances. In (Sainz et al., 2009b), analytical expressions to locate the series resonance are obtained from these admittances. Finally, the analytical expressions developed in (Sainz et al., 2009b) to predict resonance frequencies are expanded in (Sainz et al., in press) to consider the influence of Steinmetz circuit capacitor changes with respect to its design value.

This chapter, building on work developed in the previous references, not only summarizes the above research but also unifies the study of both resonances, providing an expression unique to their location. The proposed expression is the same as in the series resonance case, but substantially improves those obtained in the parallel resonance case. Moreover, the previous studies are completed with the analysis of the impact of the Steinmetz circuit inductor resistance on the resonance. This resistance, as well as damping the impedance values, shifts the resonance frequency because it influences Steinmetz circuit design (Sainz & Riera, submitted for publication). A sensitivity analysis of all variables involved in the location of the parallel and series resonance is also included. The chapter ends with several experimental tests to validate the proposed expression and several examples of its application.

## 2. Balancing ac traction systems with the Steinmetz circuit

Fig. 1a shows one of the most widely used connection schemes of ac traction systems, where the railroad substation is formed by a single-phase transformer feeding the traction load from the utility power supply system. As the railroad substation is a single-phase load which may lead to unbalanced utility supply voltages, several methods have been proposed to reduce unbalance (Chen, 1994; Hill, 1994), such as feeding railroad substations at different phases alternatively, and using special transformer connections (e.g. Scott-connection), SVCs or external balancing equipment. To simplify the study of these methods, the single-phase transformer is commonly considered ideal and the traction load is represented by its equivalent inductive impedance, $\underline{Z}_L = R_L + jX_L$, obtained from its power demand at the fundamental frequency, Fig. 1b (Arendse & Atkinson-Hope, 2010; Barnes & Wong, 1991; Chen, 1994; Mayer & Kropik, 2005; Qingzhu et al., 2010a, 2010b). According to Fig. 1c, external balancing equipment consists in the delta connection of reactances (usually an inductor $\underline{Z}_1$ and a capacitor $\underline{Z}_2$) with the single-phase load representing the railroad substation in order to load the network with balanced currents. This circuit, which is known as Steinmetz circuit (ABB Power Transmission, n.d.; Barnes & Wong, 1991; Mayer & Kropik, 2005), is not the most common balancing method in traction systems but it is also used in industrial high-power single-phase loads and electrothermal appliances (Chicco et al., 2009; Chindris et. al., 2002; Mayer & Kropik, 2005).

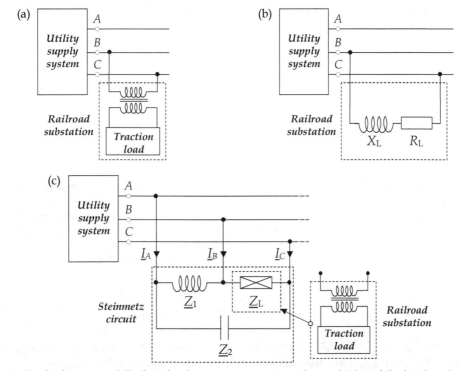

Fig. 1. Studied system: a) Railroad substation connection scheme. b) Simplified railroad substation circuit. c) Steinmetz circuit.

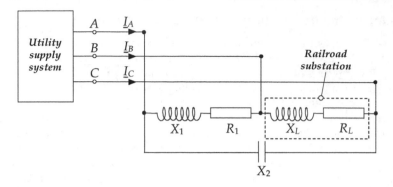

Fig. 2. Detailed Steinmetz circuit.

Fig. 2 shows the Steinmetz circuit in detail. The inductor is represented with its associated resistance, $\underline{Z}_1 = R_1 + jX_1$, while the capacitor is considered ideal, $\underline{Z}_2 = -jX_2$. Steinmetz circuit design aims to determine the reactances $X_1$ and $X_2$ to balance the currents consumed by the railroad substation. Thus, the design value of the symmetrizing reactive elements is obtained by forcing the current unbalance factor of the three-phase fundamental currents consumed by the Steinmetz circuit ($\underline{I}_A$, $\underline{I}_B$, $\underline{I}_C$) to be zero. Balanced supply voltages and the pure Steinmetz circuit inductor (i.e., $R_1 = 0$) are usually considered in Steinmetz circuit design, and the values of the reactances can be obtained from the following approximated expressions (Sainz et al., 2005):

$$X_{1,\mathrm{apr}}(R_L, \lambda_L) = \frac{\sqrt{3}R_L}{\lambda_L^2\left(1+\sqrt{3}\tau_L\right)} \quad ; \quad X_{2,\mathrm{apr}}(R_L, \lambda_L) = \frac{\sqrt{3}R_L}{\lambda_L^2\left(1-\sqrt{3}\tau_L\right)}, \quad (1)$$

where

$$\tau_L = \frac{X_L}{R_L} = \frac{\sqrt{1-\lambda_L^2}}{\lambda_L}, \quad (2)$$

and $\lambda_L = R_L/|\underline{Z}_L|$ and $|\underline{Z}_L|$ are the displacement power factor and the magnitude of the single-phase load at fundamental frequency, respectively.

In (Mayer & Kropik, 2005), the resistance of the Steinmetz inductor is considered and the symmetrizing reactance values are obtained by optimization methods. However, no analytical expressions for the reactances are provided. In (Sainz & Riera, submitted for publication), the following analytical expressions have recently been deduced

$$X_1(R_L, \lambda_L, \tau_1) = \frac{R_L\lambda_1^2\left(\sqrt{3}-\tau_1\right)}{\lambda_L^2\tau_1^2\left(1+\sqrt{3}\tau_L\right)} \quad ; \quad X_2(R_L, \lambda_L, \tau_1) = \frac{R_L\left(\sqrt{3}-\tau_1\right)}{\lambda_L^2\left\{\left(1-\sqrt{3}\tau_L\right)-\tau_1\left(\sqrt{3}+\tau_L\right)\right\}}, \quad (3)$$

where $\tau_1 = R_1/X_1 = \lambda_1/(1-\lambda_1^2)^{1/2}$ is the R/X ratio of the Steinmetz circuit inductor, and $\lambda_1 = R_1/|\underline{Z}_1|$ and $|\underline{Z}_1|$ are the displacement power factor and the magnitude of the Steinmetz circuit inductor at the fundamental frequency, respectively. It must be noted that (1) can be derived from (3) by imposing $\tau_1 = 0$ (and therefore $\lambda_1/\tau_1 = 1$). The Steinmetz circuit under study (with an inductor $X_1$ and a capacitor $X_2$) turns out to be possible only when $X_1$ and $X_2$ values are positive. Thus, according to (Sainz & Riera, submitted for publication), $X_1$ is always positive while $X_2$ is only positive when the displacement power factor of the single-phase load satisfies the condition

$$1 \geq \lambda_L \geq \lambda_{LC} = \frac{\tau_1 + \sqrt{3}}{2\sqrt{1 + \tau_1^2}}, \tag{4}$$

where the typical limit $\lambda_{LC} = (\sqrt{3})/2$ can be obtained from (4) by imposing $\tau_1 = 0$ (Jordi et al., 2002; Sainz & Riera, submitted for publication).

Supply voltage unbalance is considered in (Qingzhu et al., 2010a, 2010b) by applying optimization techniques for Steinmetz circuit design, and in (Jordi et al., 2002; Sainz & Riera, submitted for publication) by obtaining analytical expressions for the symmetrizing reactances. However, the supply voltage balance hypothesis is not as critical as the pure Steinmetz circuit inductor hypothesis. Harmonics are not considered in the literature in Steinmetz circuit design and the reactances are determined from the fundamental waveform component with the previous expressions. Nevertheless, Steinmetz circuit performance in the presence of waveform distortion is analyzed in (Arendse & Atkinson-Hope, 2010; Chicco et al., 2009). Several indicators defined in the framework of the symmetrical components are proposed to explain the properties of the Steinmetz circuit under waveform distortion.

The introduction of thyristor-controlled reactive elements due to the recent development of power electronics and the use of step variable capacitor banks allow varying the Steinmetz circuit reactances in order to compensate for the usual single-phase load fluctuations, (Barnes & Wong, 1991; Chindris et al., 2002). However, the previous design expressions must be considered in current dynamic symmetrization and the power signals are then treated by the controllers in accordance with the Steinmetz procedure for load balancing (ABB Power Transmission, n.d.; Lee & Wu, 1993; Qingzhu et al., 2010a, 2010b).

## 3. Steinmetz circuit impact on power system harmonic response

The power system harmonic response in the presence of the Steinmetz circuit is analyzed from Fig. 3. Two sources of harmonic disturbances can be present in this system: a three-phase non-linear load injecting harmonic currents into the system or a harmonic-polluted utility supply system. In the former, the parallel resonance may affect power quality because harmonic voltages due to injected harmonic currents can be magnified. In the latter, series resonance may affect power quality because consumed harmonic currents due to background voltage distortion can also be magnified. Therefore, the system harmonic response depends on the equivalent harmonic impedance or admittance "observed" from the three-phase load or the utility supply system, respectively. This chapter, building on work developed in (Sainz et al., 2007, in press), summarizes the above research on parallel and series resonance location and unifies this study. It provides an expression unique to the location of the parallel and series resonance considering the Steinmetz circuit inductor resistance.

In Fig. 3, the impedances $\underline{Z}_{Lk} = R_L + jkX_L$, $\underline{Z}_{1k} = R_1 + jkX_1$ and $\underline{Z}_{2k} = -jX_2/k$ represent the single-phase load, the inductor and the capacitor of the Steinmetz circuit at the fundamental ($k = 1$) and harmonic frequencies ($k > 1$). Note that impedances $\underline{Z}_{L1}$, $\underline{Z}_{11}$ and $\underline{Z}_{21}$ correspond to impedances $\underline{Z}_L$, $\underline{Z}_1$ and $\underline{Z}_2$ in Section 2, respectively. Moreover, parameter $d_C$ is introduced in the study representing the degree of the Steinmetz circuit capacitor degradation from its design value [(1) or (3)]. Thus, the capacitor value considered in the harmonic study is $d_C \cdot C$, i.e. $-j1/(d_C \cdot C \omega_1 \cdot k) = -j \cdot (X_2/k)/d_C = \underline{Z}_{2k}/d_C$ where $\omega_1 = 2\pi \cdot f_1$ and $f_1$ is the fundamental frequency of the supply voltage. This parameter allows examining the impact of the capacitor bank degradation caused by damage in the capacitors or in their fuses on the power system harmonic response. If $d_C = 1$, the capacitor has the design value [(1) or (3)] whereas if $d_C < 1$, the capacitor value is lower than the design value.

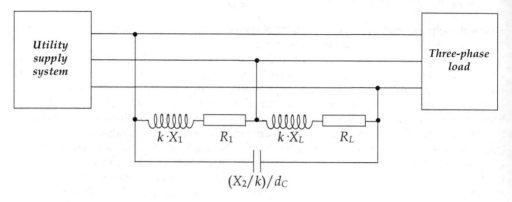

Fig. 3. Power system harmonic analysis in the presence of the Steinmetz circuit.

### 3.1 Study of the parallel resonance

This Section examines the harmonic response of the system "observed" from the three-phase load. It implies analyzing the passive set formed by the utility supply system and the Steinmetz circuit (see Fig. 4). The system harmonic behavior is characterized by the equivalent harmonic impedance matrix, $\mathbf{Z}_{\mathbf{Busk}}$, which relates the $k$th harmonic three-phase voltages and currents at the three-phase load node, $\underline{\mathbf{V}}_k = [\underline{V}_{Ak} \; \underline{V}_{Bk} \; \underline{V}_{Ck}]^T$ and $\underline{\mathbf{I}}_k = [\underline{I}_{Ak} \; \underline{I}_{Bk} \; \underline{I}_{Ck}]^T$. Thus, considering point $N$ in Fig. 4 as the reference bus, this behavior can be characterized by the voltage node method,

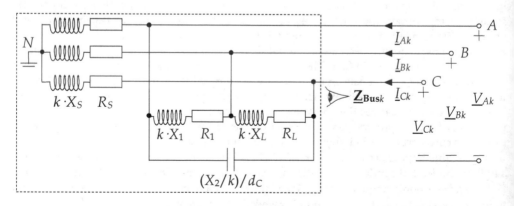

Fig. 4. Study of the parallel resonance in the presence of the Steinmetz circuit.

$$\mathbf{V}_k = \mathbf{Z}_{\text{Bus}k} \cdot \mathbf{I}_k \quad \Rightarrow \quad \begin{bmatrix} \underline{V}_{Ak} \\ \underline{V}_{Bk} \\ \underline{V}_{Ck} \end{bmatrix} = \begin{bmatrix} \underline{Z}_{AAk} & \underline{Z}_{ABk} & \underline{Z}_{ACk} \\ \underline{Z}_{BAk} & \underline{Z}_{BBk} & \underline{Z}_{BCk} \\ \underline{Z}_{CAk} & \underline{Z}_{CBk} & \underline{Z}_{CCk} \end{bmatrix} \cdot \begin{bmatrix} \underline{I}_{Ak} \\ \underline{I}_{Bk} \\ \underline{I}_{Ck} \end{bmatrix}$$

$$= \begin{bmatrix} \underline{Y}_{Sk} + \underline{Y}_{1k} + d_C \underline{Y}_{2k} & -\underline{Y}_{1k} & -d_C \underline{Y}_{2k} \\ -\underline{Y}_{1k} & \underline{Y}_{Sk} + \underline{Y}_{1k} + \underline{Y}_{Lk} & -\underline{Y}_{Lk} \\ -d_C \underline{Y}_{2k} & -\underline{Y}_{Lk} & \underline{Y}_{Sk} + d_C \underline{Y}_{2k} + \underline{Y}_{Lk} \end{bmatrix}^{-1} \cdot \begin{bmatrix} \underline{I}_{Ak} \\ \underline{I}_{Bk} \\ \underline{I}_{Ck} \end{bmatrix},$$

(5)

where

- $\underline{Y}_{Sk} = \underline{Z}_{Sk}^{-1} = (R_S + jkX_S)^{-1}$ corresponds to the admittance of the power supply system, which includes the impedance of the power supply network, the short-circuit impedance of the three-phase transformer and the impedance of the overhead lines feeding the Steinmetz circuit and the three-phase linear load.
- $\underline{Y}_{Lk}$, $\underline{Y}_{1k}$ and $d_C \cdot \underline{Y}_{2k}$ correspond to the admittances of the Steinmetz circuit components (i.e., the inverse of the impedances $\underline{Z}_{Lk}$, $\underline{Z}_{1k}$ and $\underline{Z}_{2k}/d_C$ in Section 3, respectively).

It can be observed that the diagonal and non-diagonal impedances of the harmonic impedance matrix $\mathbf{Z}_{\text{Bus}k}$ (i.e., $\underline{Z}_{AAk}$ to $\underline{Z}_{CCk}$) directly characterize the system harmonic behavior. Diagonal impedances are known as phase driving point impedances (Task force on Harmonic Modeling and Simulation, 1996) since they allow determining the contribution of the harmonic currents injected into any phase $F$ ($\underline{I}_{Fk}$) to the harmonic voltage of this phase ($\underline{V}_{Fk}$). Non-diagonal impedances are the equivalent impedances between a phase and the rest of the phases since they allow determining the contribution of the harmonic currents injected into any phase $F$ ($\underline{I}_{Fk}$) to the harmonic voltage of any other phase $G$ ($\underline{V}_{Gk}$, with $G \neq F$). Thus, the calculation of both sets of impedances is necessary because a resonance in either of them could cause a high level of distortion in the corresponding voltages and damage harmonic power quality.

As an example, a network as that in Fig. 4 was constructed in the laboratory and its harmonic response (i.e., $\mathbf{Z}_{\text{Bus}k}$ matrix) was measured with the following per unit data ($U_B = 100$ V and $S_B = 500$ VA) and considering two cases ($d_C = 1$ and 0.5):

- Supply system: $\underline{Z}_{S1} = 0.022 + j0.049$ pu.
- Railroad substation: $R_L = 1.341$ pu, $\lambda_L = 1.0$.
- External balancing equipment: According to (1) and (3), two pairs of reactances were connected with the railroad substation, namely $X_{1, \text{apr}} = 2.323$ pu and $X_{2, \text{apr}} = 2.323$ pu and $X_1 = 2.234$ pu and $X_2 = 2.503$ pu. The former was calculated by neglecting the inductor resistance (1) and the latter was calculated by considering the actual value of this resistance (3) ($R_1 \approx 0.1342$ pu, and therefore $\tau_1 \approx 0.1341/2.234 = 0.06$).

Considering that the system fundamental frequency is 50 Hz, the measurements of the $\mathbf{Z}_{\text{Bus}k}$ impedance magnitudes (i.e., $|\underline{Z}_{AAk}|$ to $|\underline{Z}_{CCk}|$) with $X_{1, \text{apr}}$ and $X_{2, \text{apr}}$ are plotted in Fig. 5 for both cases ($d_C = 1$ in solid lines and $d_C = 0.5$ in broken lines). It can be noticed that

- The connection of the Steinmetz circuit causes a parallel resonance in the $\mathbf{Z}_{\text{Bus}k}$ impedances that occurs in phases $A$ and $C$, between which the capacitor is connected, and is located nearly at the same harmonic for all the impedances (labeled as $k_{p, \text{meas}}$). This asymmetrical resonant behavior has an asymmetrical effect on the harmonic voltages (i.e., phases $A$ and $C$ have the highest harmonic impedance, and therefore the highest harmonic voltages.)

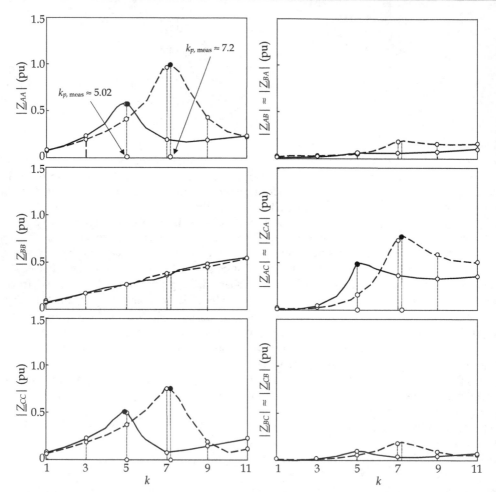

Fig. 5. Measured impedance - frequency matrix in the presence of the Steinmetz circuit with $X_{1,\,apr} = X_{2,\,apr} = 2.323$ pu (solid line: $d_C = 1$; broken line: $d_C = 0.5$).

- In the case of $d_C = 1$ (in solid lines), the connection of the Steinmetz circuit causes a parallel resonance measured close to the fifth harmonic ($k_{p,\,meas} \approx 251/50 = 5.02$, where 251 Hz is the frequency of the measured parallel resonance.)
- If the Steinmetz circuit suffers capacitor bank degradation, the parallel resonance is shifted to higher frequencies. In the example, a 50% capacitor loss (i.e., $d_C = 0.5$ in broken lines) shifts the parallel resonance close to the seventh harmonic ($k_{s,\,meas} \approx 360/50 = 7.2$, where 360 Hz is the frequency of the measured parallel resonance.)

The measurements of the $\mathbf{Z}_{\mathbf{Bus}k}$ impedance magnitudes (i.e. $|\underline{Z}_{AAk}|$ to $|\underline{Z}_{CCk}|$) with $X_1$ and $X_2$ are not plotted for space reasons. In this case, the parallel resonance shifts to $k_{p,\,meas} \approx 5.22$ ($d_C = 1$) and $k_{p,\,meas} \approx 7.43$ ($d_C = 0.5$) but the general conclusions of the $X_{1,\,apr}$ and $X_{2,\,apr}$ case are true.

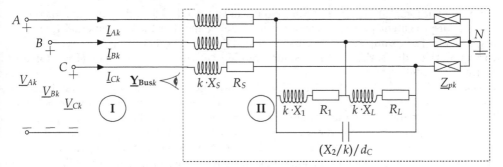

Fig. 6. Study of the series resonance in the presence of the Steinmetz circuit.

## 3.2 Study of the series resonance

This Section studies the harmonic response of the system "observed" from the utility supply system. It implies analyzing the passive set formed by the supply system impedances, the Steinmetz circuit and the three-phase load (see Fig. 6). The system harmonic behavior is characterized by the equivalent harmonic admittance matrix, $\mathbf{Y_{Busk}}$, which relates the $k^{th}$ harmonic three-phase currents and voltages at the node I, $\mathbf{I}^I_k = [\underline{I}_{Ak}\ \underline{I}_{Bk}\ \underline{I}_{Ck}]^T$ and $\mathbf{V}^I_k = [\underline{V}_{Ak}\ \underline{V}_{Bk}\ \underline{V}_{Ck}]^T$, respectively. Thus, considering point $N$ in Fig. 6 as the reference bus, this behavior can be characterized by the voltage node method,

$$\begin{bmatrix} \mathbf{I}^I_k \\ \hline 0 \end{bmatrix} = \left[ \begin{array}{c|c} \mathbf{Y}^I_k & \mathbf{Y}^{I\text{-}II}_k \\ \hline \mathbf{Y}^{II\text{-}I}_k & \mathbf{Y}^{II}_k \end{array} \right] \cdot \begin{bmatrix} \mathbf{V}^I_k \\ \hline \mathbf{V}^{II}_k \end{bmatrix}$$

$$\Rightarrow \begin{bmatrix} \underline{I}_{Ak} \\ \underline{I}_{Bk} \\ \underline{I}_{Ck} \\ \hline 0 \\ 0 \\ 0 \end{bmatrix} = \left[ \begin{array}{ccc|ccc} \underline{Y}_{Sk} & 0 & 0 & -\underline{Y}_{Sk} & 0 & 0 \\ 0 & \underline{Y}_{Sk} & 0 & 0 & -\underline{Y}_{Sk} & 0 \\ 0 & 0 & \underline{Y}_{Sk} & 0 & 0 & -\underline{Y}_{Sk} \\ \hline -\underline{Y}_{Sk} & 0 & 0 & \underline{Y}_{T1k} & -\underline{Y}_{1k} & -d_C\underline{Y}_{2k} \\ 0 & -\underline{Y}_{Sk} & 0 & -\underline{Y}_{1k} & \underline{Y}_{T2k} & -\underline{Y}_{Lk} \\ 0 & 0 & -\underline{Y}_{Sk} & -d_C\underline{Y}_{2k} & -\underline{Y}_{Lk} & \underline{Y}_{T3k} \end{array} \right] \cdot \begin{bmatrix} \underline{V}_{Ak} \\ \underline{V}_{Bk} \\ \underline{V}_{Ck} \\ \hline \underline{V}^{II}_{Ak} \\ \underline{V}^{II}_{Bk} \\ \underline{V}^{II}_{Ck} \end{bmatrix}, \quad (6)$$

and

$$\mathbf{I}^I_k = \begin{bmatrix} \underline{I}_{Ak} \\ \underline{I}_{Bk} \\ \underline{I}_{Ck} \end{bmatrix} = \left( \mathbf{Y}^I_k - \mathbf{Y}^{I\text{-}II}_k {\mathbf{Y}^{II}_k}^{-1} \mathbf{Y}^{II\text{-}I}_k \right) \mathbf{V}^I_k = \mathbf{Y_{Busk}}\, \mathbf{V}^I_k = \begin{bmatrix} \underline{Y}_{AAk} & \underline{Y}_{ABk} & \underline{Y}_{ACk} \\ \underline{Y}_{BAk} & \underline{Y}_{BBk} & \underline{Y}_{BCk} \\ \underline{Y}_{CAk} & \underline{Y}_{CBk} & \underline{Y}_{CCk} \end{bmatrix} \cdot \begin{bmatrix} \underline{V}_{Ak} \\ \underline{V}_{Bk} \\ \underline{V}_{Ck} \end{bmatrix}, \quad (7)$$

where

$$\underline{Y}_{T1k} = \underline{Y}_{Sk} + \underline{Y}_{Pk} + \underline{Y}_{1k} + d_C\underline{Y}_{2k} \quad ; \quad \underline{Y}_{T2k} = \underline{Y}_{Sk} + \underline{Y}_{Pk} + \underline{Y}_{1k} + \underline{Y}_{Lk}$$
$$\underline{Y}_{T3k} = \underline{Y}_{Sk} + \underline{Y}_{Pk} + \underline{Y}_{Lk} + d_C\underline{Y}_{2k}. \quad (8)$$

Admittances $\underline{Y}_{Sk}$, $\underline{Y}_{Lk}$, $\underline{Y}_{1k}$ and $d_C \cdot \underline{Y}_{2k}$ were already introduced in the parallel resonance location and $\underline{Y}_{Pk} = \underline{Z}_{Pk}^{-1} = g_{LM\#}(\,|\underline{Z}_{P1}|\,,\ \lambda_P,\ k)$ is the three-phase load admittance. The function $g_{LM\#}(\cdot)$ represents the admittance expressions of the load models 1 to 7 proposed in (Task force on Harmonic Modeling and Simulation, 2003), and $|\underline{Z}_{P1}|$ and $\lambda_P$ are the magnitude and the displacement power factor of the load impedances at the fundamental frequency, respectively. For example, the expression $g_{LM1}(|\underline{Z}_{P1}|, \lambda_P, k) = 1/\{|\underline{Z}_{P1}| \cdot (\lambda_P + jk(1 - \lambda_P^2)^{1/2})\}$ corresponds to the series R-L impedance model, i.e. the model LM1 in (Task force on Harmonic Modeling and Simulation, 2003).

It can be observed that the diagonal and non-diagonal admittances of the harmonic admittance matrix $\mathbf{Y}_{Busk}$ (i.e., $\underline{Y}_{AAk}$ to $\underline{Y}_{CCk}$) directly characterize system harmonic behavior. Diagonal admittances allow determining the contribution of the harmonic voltages at any phase $F$ ($\underline{V}_{Fk}$) to the harmonic currents consumed at this phase ($\underline{I}_{Fk}$). Non-diagonal admittances allow determining the contribution of the harmonic voltages at any phase $F$ ($\underline{V}_{Fk}$) to the harmonic currents consumed at any other phase $G$ ($\underline{I}_{Gk}$, with $G \neq F$). Thus, the calculation of both sets of admittances is necessary because this resonance could lead to a high value of the admittance magnitude, magnify the harmonic currents consumed in the presence of background voltage distortion and damage harmonic power quality.

As an example, a network as that in Fig. 6 was constructed in the laboratory and its harmonic response (i.e., $\mathbf{Y}_{Busk}$ matrix) was measured with the following per unit data ($U_B = 100$ V and $S_B = 500$ VA) and considering two cases ($d_C = 1$ and 0.5):

- Supply system: $\underline{Z}_{S1} = 0.076 + j0.154$ pu.
- Railroad substation: $R_L = 1.464$ pu, $\lambda_L = 0.95$.
- External balancing equipment: According to (1) and (3), two pairs of reactances were connected with the railroad substation, namely $X_{1, apr} = 1.790$ pu and $X_{2, apr} = 6.523$ pu and $X_1 = 1.698$ pu and $X_2 = 10.03$ pu. The former was calculated by neglecting the inductor resistance (1) and the latter was calculated by considering the actual value of this resistance (3) ($R_1 \approx 0.1342$ pu, and therefore $\tau_1 \approx 0.1341/1.698 = 0.079$).
- Three-phase load: Grounded wye series R-L impedances with $|\underline{Z}_{P1}| = 30.788$ pu and $\lambda_P = 0.95$ are connected, i.e. the three-phase load model LM1 in (Task force on Harmonic Modeling and Simulation, 2003).

Considering that the system fundamental frequency is 50 Hz, the measurements of the $\mathbf{Y}_{Busk}$ admittance magnitudes (i.e. $|\underline{Y}_{AAk}|$ to $|\underline{Y}_{CCk}|$) with $X_{1, apr}$ and $X_{2, apr}$ are plotted in Fig. 7 for both cases ($d_C = 1$ in solid lines and $d_C = 0.5$ in broken lines). It can be noted that

- The connection of the Steinmetz circuit causes a series resonance in the $\mathbf{Y}_{Busk}$ admittances that occurs in phases $A$ and $C$, between which the capacitor is connected, and is located nearly at the same harmonic for all the admittances (labeled as $k_{s, meas}$). This asymmetrical resonant behavior has an asymmetrical effect on the harmonic consumed currents (i.e., phases $A$ and $C$ have the highest harmonic admittance, and therefore the highest harmonic consumed currents.)
- In the case of $d_C = 1$ (in solid lines), the connection of the Steinmetz circuit causes a series resonance measured close to the fifth harmonic ($k_{s, meas} \approx 255/50 = 5.1$, where 255 Hz is the frequency of the measured series resonance.)
- If the Steinmetz circuit suffers capacitor bank degradation, the series resonance is shifted to higher frequencies. In the example, a 50% capacitor loss (i.e., $d_C = 0.5$ in broken lines) shifts the series resonance close to the seventh harmonic ($k_{s, meas} \approx 365/50 = 7.3$, where 365 Hz is the frequency of the measured series resonance.)

The measurements of the $\mathbf{Y}_{Busk}$ admittance magnitudes (i.e., $|\underline{Y}_{AAk}|$ to $|\underline{Y}_{CCk}|$) with $X_1$ and $X_2$ are not plotted for space reasons. In this case, the series resonance shifts to $k_{s, meas} \approx 6.31$ ($d_C = 1$) and $k_{s, meas} \approx 8.83$ ($d_C = 0.5$) but the general conclusions of the $X_{1, apr}$ and $X_{2, apr}$ case are true.

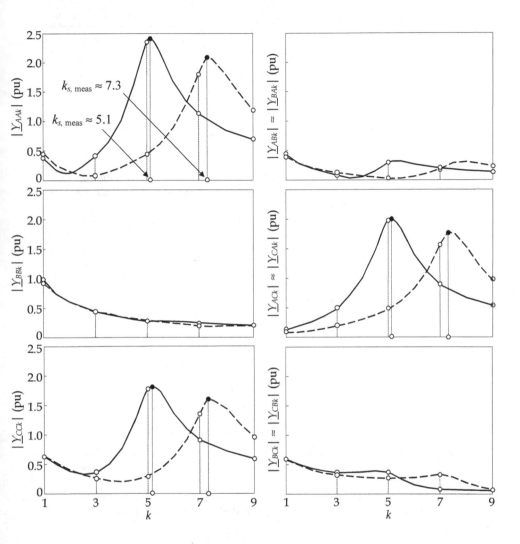

Fig. 7. Measured admittance - frequency matrix in the presence of the Steinmetz circuit with $X_{1, apr}$ = 1.790 pu and $X_{2, apr}$ = 6.523 pu (solid line: $d_C$= 1; broken line: $d_C$ = 0.5).

## 4. Analytical study of power system harmonic response

In this Section, the magnitudes of the most critical $\mathbf{Z}_{Busk}$ impedances (i.e., $|\underline{Z}_{AAk}|$, $|\underline{Z}_{CCk}|$, $|\underline{Z}_{ACk}|$ and $|\underline{Z}_{CAk}|$) and $\mathbf{Y}_{Busk}$ admittances (i.e., $|\underline{Y}_{AAk}|$, $|\underline{Y}_{CCk}|$, $|\underline{Y}_{ACk}|$ and $|\underline{Y}_{CAk}|$) are analytically studied in order to locate the parallel and series resonance, respectively.

### 4.1 Power system harmonic characterization

The most critical $\mathbf{Z}_{Busk}$ impedances are obtained from (5):

$$\left|\underline{Z}_{AAk}\right| = \left|\frac{\underline{Y}_{Sk}^2 + \underline{Y}_{Sk}(\underline{Y}_{St\pm1k} + \underline{Y}_{Lk}) + \underline{Y}_{St\pm2k}}{\underline{Y}_{Sk}\underline{D}_k}\right| \quad ; \quad \left|\underline{Z}_{CCk}\right| = \left|\frac{\underline{Y}_{Sk}^2 + \underline{Y}_{Sk}(\underline{Y}_{St\pm1k} + \underline{Y}_{1k}) + \underline{Y}_{St\pm2k}}{\underline{Y}_{Sk}\underline{D}_k}\right|$$

$$\left|\underline{Z}_{ACk}\right| = \left|\underline{Z}_{CAk}\right| = \left|\frac{\underline{Y}_{St\pm2k} + d_C \underline{Y}_{2k}\underline{Y}_{Sk}}{\underline{Y}_{Sk}\underline{D}_k}\right|, \tag{9}$$

where

$$\underline{Y}_{St\pm1k} = \underline{Y}_{1k} + d_C\underline{Y}_{2k} + \underline{Y}_{Lk} \quad ; \quad \underline{Y}_{St\pm2k} = \underline{Y}_{Lk}\underline{Y}_{1k} + \underline{Y}_{Lk}d_C\underline{Y}_{2k} + \underline{Y}_{1k}d_C\underline{Y}_{2k}$$

$$\underline{D}_k = \underline{Y}_{Sk}^2 + 2\underline{Y}_{Sk}\underline{Y}_{St\pm1k} + 3\cdot\underline{Y}_{St\pm2k}. \tag{10}$$

The harmonic of the parallel resonance numerically obtained as the maximum value of the above impedance magnitudes is located nearly at the same harmonic for all the impedances (Sainz et al., 2007) and labeled as $k_{p,n}$.

The most critical $\underline{Y}_{Busk}$ admittances are obtained from (7):

$$\left|\underline{Y}_{AAk}\right| = \left|\frac{\underline{Y}_{Sk}(\underline{Y}_{Sk}\underline{N}_{AAk} + \underline{Y}_{Pk}\underline{N}_{AAk}^{(P)})}{\underline{Y}_{Sk}\underline{D}_k + \underline{Y}_{Pk}\cdot\underline{D}_k^{(P)}}\right| \quad ; \quad \left|\underline{Y}_{CCk}\right| = \left|\frac{\underline{Y}_{Sk}(\underline{Y}_{Sk}\underline{N}_{CCk} + \underline{Y}_{Pk}\underline{N}_{CCk}^{(P)})}{\underline{Y}_{Sk}\underline{D}_k + \underline{Y}_{Pk}\cdot\underline{D}_k^{(P)}}\right|$$

$$\left|\underline{Y}_{ACk}\right| = \left|\underline{Y}_{CAk}\right| = \left|\frac{\underline{Y}_{Sk}^2(\underline{Y}_{Sk}d_C\underline{Y}_{2k} + \underline{Y}_{St\pm2k} + \underline{Y}_{Pk}d_C\underline{Y}_{2k})}{\underline{Y}_{Sk}\underline{D}_k + \underline{Y}_{Pk}\cdot\underline{D}_k^{(P)}}\right|, \tag{11}$$

where

$$\underline{N}_{AAk} = \underline{Y}_{Sk}(\underline{Y}_{1k} + d_C\underline{Y}_{2k}) + 2\cdot\underline{Y}_{St\pm2k} \quad ; \quad \underline{N}_{CCk} = \underline{Y}_{Sk}(\underline{Y}_{Lk} + d_C\underline{Y}_{2k}) + 2\cdot\underline{Y}_{St\pm2k}$$

$$\underline{N}_{AAk}^{(P)} = \underline{Y}_{Pk}^2 + 2\cdot(\underline{Y}_{Sk} + \underline{Y}_{St\pm1k})\underline{Y}_{Pk} + 3\cdot(\underline{Y}_{St\pm2k} + \underline{Y}_{Sk}(\underline{Y}_{1k} + d_C\underline{Y}_{2k})) + \underline{Y}_{Sk}(\underline{Y}_{Sk} + 2\cdot\underline{Y}_{Lk})$$

$$\underline{N}_{CCk}^{(P)} = \underline{Y}_{Pk}^2 + 2\cdot(\underline{Y}_{Sk} + \underline{Y}_{St\pm1k})\underline{Y}_{Pk} + 3\cdot(\underline{Y}_{St\pm2k} + \underline{Y}_{Sk}(\underline{Y}_{Lk} + d_C\underline{Y}_{2k})) + \underline{Y}_{Sk}(\underline{Y}_{Sk} + 2\cdot\underline{Y}_{1k})$$

$$\underline{D}_k^{(P)} = \underline{Y}_{Pk}^2 + (3\underline{Y}_{Sk} + 2\cdot\underline{Y}_{St\pm1k})\underline{Y}_{Pk} + 3\cdot(\underline{Y}_{Sk}^2 + \underline{Y}_{St\pm2k}) + 4\underline{Y}_{Sk}\underline{Y}_{St\pm1k}. \tag{12}$$

The harmonic of the series resonance numerically obtained as the maximum value of the above admittance magnitudes is located nearly at the same harmonic for all the admittances (Sainz et al., in press) and labeled as $k_{s,n}$.

Since the expressions of the $\underline{Y}_{Busk}$ admittances (11) are too complicated to be analytically analyzed, the admittance $\underline{Y}_{Pk}$ is not considered in their determination (i.e., $\underline{Y}_{Pk} = 0$), and they are approximated to

$$\left|\underline{Y}_{AAk}\right| \approx \left|\underline{Y}_{AAk,apx}\right| = \left|\frac{\underline{Y}_{Sk}\underline{N}_{AAk}}{\underline{D}_k}\right| \quad ; \quad \left|\underline{Y}_{CCk}\right| \approx \left|\underline{Y}_{CCk,apx}\right| = \left|\frac{\underline{Y}_{Sk}\underline{N}_{CCk}}{\underline{D}_k}\right|$$

$$\left|\underline{Y}_{ACk}\right| = \left|\underline{Y}_{CAk}\right| \approx \left|\underline{Y}_{ACk,apx}\right| = \left|\underline{Y}_{ACk,apx}\right| = \left|\frac{\underline{Y}_{Sk}(\underline{Y}_{Sk}d_C\underline{Y}_{2k} + \underline{Y}_{St\pm2k})}{\underline{D}_k}\right|. \tag{13}$$

This approximation is based on the fact that, the three-phase load does not influence the series resonance significantly if large enough, (Sainz et al., 2009b). The harmonic of the series resonance numerically obtained as the maximum value of the above admittance magnitudes is located nearly at the same harmonic for all the admittances (Sainz et al., in press) and labeled as $k_{s,napx}$.

It must be noted that (9) and (13) depend on the supply system and the Steinmetz circuit admittances (i.e., $\underline{Y}_{Sk}$, $\underline{Y}_{Lk}$, $\underline{Y}_{1k}$ and $d_C \cdot \underline{Y}_{2k}$) and (11) depends on the previous admittances and three-phase load admittances (i.e., $\underline{Y}_{Sk}$, $\underline{Y}_{Lk}$, $\underline{Y}_{1k}$, $d_C \cdot \underline{Y}_{2k}$ and $\underline{Y}_{Pk}$). In the study, these admittances are written as

$$\underline{Y}_{Sk} \approx -j\frac{1}{kX_S} \quad ; \quad \underline{Y}_{Lk} = \frac{1}{R_L(1+jk\tau_L)} \quad ; \quad \underline{Y}_{1k} \approx \frac{1}{jk \cdot X_1} = -j\frac{1}{k}\left(\frac{\lambda_L \tau_1}{\lambda_1}\right)^2 \frac{1+\sqrt{3}\tau_L}{R_L(\sqrt{3}-\tau_1)}$$

$$d_C \underline{Y}_{2k} = \frac{d_C}{-jX_2/k} = jk\frac{d_C \lambda_L^2\left\{\left(1-\sqrt{3}\tau_L\right)-\tau_1\left(\sqrt{3}+\tau_L\right)\right\}}{R_L\left(\sqrt{3}-\tau_1\right)} \quad ; \quad \underline{Y}_{Pk} = g_{LM\#}\left(\left|\underline{Z}_{P1}\right|, \lambda_p, k\right),$$

(14)

where (3) is used to obtain the Steinmetz circuit components $X_1$ and $X_2$, and $g_{LM\#}(\cdot)$ represents the three-phase load admittance models 1 to 7 proposed in (Task force on Harmonic Modeling and Simulation, 2003). Thus, it is observed that the $\underline{Z}_{Busk}$ impedances and $\underline{Y}_{Busk}$ admittances are functions of eight variables, namely

- the harmonic order $k$,
- the supply system fundamental reactance $X_S$,
- the single-phase load resistance $R_L$,
- the parameter $\tau_L$, i.e. the single-phase load fundamental displacement power factor $\lambda_L$ (2),
- the R/X ratio of the Steinmetz circuit inductor $\tau_1$,
- the degradation parameter $d_C$,
- the magnitude of the linear load fundamental impedance $|\underline{Z}_{P1}|$ and
- the linear load fundamental displacement power factor $\lambda_P$.

It is worth pointing out that the resistance of the supply system is neglected (i.e., $\underline{Z}_{Sk} = R_S + jkX_S \approx jkX_S$) and the resistance of the Steinmetz circuit is only considered in the inductor design [i.e., $\underline{Z}_{1k} = R_1 + jkX_1 \approx jkX_1$ and $\underline{Z}_{2k} = -j \cdot X_2/k$, with $X_1$ and $X_2$ obtained from (3)]. This is because the real part of these impedances does not modify the series resonance frequency significantly (Sainz et al., 2007, 2009a) while the impact of $R_1$ on Steinmetz circuit design modifies the resonance. This influence is not considered in the previous harmonic response studies.

In order to reduce the above number of variables, the $\underline{Z}_{Busk}$ impedances and $\underline{Y}_{Busk}$ admittances are normalized with respect to the supply system fundamental reactance $X_S$. For example, the normalized magnitudes $|\underline{Z}_{AAk}|_N$ and $|\underline{Y}_{AAk}|_N$ can be expressed from (9) and (11) as

$$\left|\underline{Z}_{AAk}\right|_N = \frac{\left|\underline{Z}_{AAk}\right|}{X_S} = \frac{1}{X_S}\left|\frac{X_S^2\left\{\underline{Y}_{Sk}^2 + \underline{Y}_{Sk}(\underline{Y}_{St z1k} + \underline{Y}_{Lk}) + \underline{Y}_{St z2k}\right\}}{X_S^2 \underline{Y}_{Sk} D_k}\right|,$$

$$\left|\underline{Y}_{AAk}\right|_N = X_S\left|\underline{Y}_{AAk}\right|_N = X_S\left|\frac{X_S^3 \underline{Y}_{Sk}(\underline{Y}_{Sk} N_{AAk} + \underline{Y}_{Pk} N_{AAk}^{(P)})}{X_S^3\left\{\underline{Y}_{Sk} D_k + \underline{Y}_{Pk} \cdot D_k^{(P)}\right\}}\right|,$$

(15)

where (15) has only terms $X_S \cdot \underline{Y}_{Sk}$, $X_S \cdot \underline{Y}_{Lk}$, $X_S \cdot \underline{Y}_{1k}$, $X_S \cdot d_C \cdot \underline{Y}_{2k}$ and $X_S \cdot \underline{Y}_{Pk}$, which can be rewritten from (14) as

$$X_S \underline{Y}_{Sk} = -j\frac{1}{k} \quad , \quad X_S \underline{Y}_{Lk} = \frac{1}{r_L(1+jk\tau_L)} \quad , \quad X_S \underline{Y}_{Pk} = g_{\text{LM\#, N}}(z_P, \lambda_P, k)$$

$$X_S \underline{Y}_{1k} = \frac{X_S}{jk \cdot X_1} = -j\frac{1}{k \cdot x_1} = -j\frac{1}{k}\left(\frac{\lambda_L \tau_1}{\lambda_1}\right)^2 \frac{1+\sqrt{3}\tau_L}{r_L(\sqrt{3}-\tau_1)}, \quad (16)$$

$$X_S d_C \underline{Y}_{2k} = \frac{X_S d_C}{-jX_2/k} = jk\frac{d_C}{x_2} = jk\frac{d_C \lambda_L^2\{(1-\sqrt{3}\tau_L)-\tau_1(\sqrt{3}+\tau_L)\}}{r_L(\sqrt{3}-\tau_1)},$$

where $g_{\text{LM\#, N}}(\cdot)$ represents the normalized expressions of the three-phase load models 1 to 7 proposed in (Task force on Harmonic Modeling and Simulation, 2003), $r_L = R_L / X_S$ and $z_P = |\underline{Z}_{P1}| / X_S$. The normalized expressions $g_{\text{LM\#, N}}(\cdot)$ are obtained and presented in (Sainz et al., 2009a) but are not included in the present text for space reasons. As an example, the normalized expression of model LM1 in (Task force on Harmonic Modeling and Simulation, 2003) is $g_{\text{LM1, N}}(z_P, \lambda_P, k) = 1/\{z_P \cdot (\lambda_P + jk(1-\lambda_P^2)^{1/2})\}$ (Sainz et al., 2009a).

From (15), it is interesting to note that the normalization does not modify the parallel and series resonance ($k_{p, n}$, $k_{s, n}$ and $k_{s, napx}$), but the number of variables of the normalized $\underline{Z}_{\text{Busk}}$ impedances and $\underline{Y}_{\text{Busk}}$ admittances are reduced to seven (16), i.e.,

- the harmonic order $k$,
- the ratio of the single-phase load resistance to the supply system fundamental reactance $r_L = R_L / X_S$,
- the parameter $\tau_L$, i.e. the single-phase load fundamental displacement power factor $\lambda_L$ (2),
- the R/X ratio of the Steinmetz circuit inductor $\tau_1$,
- the degradation parameter $d_C$,
- the ratio of the linear load fundamental impedance magnitude to the supply system fundamental reactance $z_P = |\underline{Z}_{P1}| / X_S$ and
- the linear load fundamental displacement power factor $\lambda_P$.

Moreover, the usual ranges of values of these variables can be obtained by relating them with known parameters to study resonances under power system operating conditions. Thus, the power system harmonic response is analyzed for the following variable ranges:

- Harmonic: $k = (1, ..., 15)$.
- Single-phase load: $r_L = (5, ..., 1000)$ and $\lambda_L = (0.9, ..., 1)$.
- Steinmetz circuit inductor: $\tau_1 = (0, ..., 0.5)$.
- Degradation parameter: $d_C = (0.25, ..., 1)$.
- Linear load: $z_P = (5, ..., 1000)$ and $\lambda_P = (0.9, ..., 1)$.

The ratios $r_L$ and $z_P$ are equal to the ratios $\lambda_L \cdot S_S / S_L$ and $S_S / S_P$ (Sainz et al., 2009a), where $S_S$ is the short-circuit power at the PCC bus, $S_L$ is the apparent power of the single-phase load and $S_P$ is the apparent power of the three-phase load. Thus, the range of these ratios is determined considering the usual values of the ratios $S_S / S_L$ and $S_S / S_P$ (Chen, 1994; Chen & Kuo, 1995) and the fundamental displacement power factors $\lambda_L$ and $\lambda_P$.

In next Section, the normalized magnitudes of the most critical $\underline{Z}_{\text{Busk}}$ impedances (9) and approximated $\underline{Y}_{\text{Busk}}$ admittances (13) are analytically studied to obtain simple expressions for locating the parallel and series resonance. Thus, these expressions are functions of the following five variables only: $k$, $r_L = R_L / X_S$, $\lambda_L$, $\tau_1$ and $d_C$.

The series resonance study in the next Section is only valid for $z_P > 20$ because the approximation of not considering the three-phase load admittance (i.e., $Y_{Pk} = 0$) is based on the fact that this load does not strongly influence the series resonance if $z_P$ is above 20. Nevertheless, the magnitude of the normalized admittances at the resonance point is low for $z_P < 20$ and the consumed currents do not increase significantly (Sainz et al., 2009a).

## 4.2 Analytical location of the parallel and series resonance

It is numerically verified that the parallel and series resonance, i.e. the maximum magnitude values of the normalized $\mathbf{Z}_{Busk}$ impedances and approximated $\mathbf{Y}_{Busk}$ admittances [obtained from (9) and (13)] with respect to the harmonic $k$ respectively, coincide with the minimum value of their denominators for the whole range of system variables. Thus, from (9) and (13), these denominators can be written as

$$
\begin{aligned}
\text{Den}\left(\left|\underline{Y}_{AAk,\text{apx}}\right|_N\right) &= \text{Den}\left(\left|\underline{Y}_{CCk,\text{apx}}\right|_N\right) = k \cdot \text{Den}\left(\left|\underline{Y}_{ACk,\text{apx}}\right|_N\right) \\
&= k \cdot \text{Den}\left(\left|\underline{Z}_{AAk}\right|_N\right) = k \cdot \text{Den}\left(\left|\underline{Z}_{CCk}\right|_N\right) = k \cdot \text{Den}\left(\left|\underline{Z}_{ACk}\right|_N\right) \\
&= k\left|k(H_1 k^2 + H_2) + j \cdot (H_3 k^2 + H_4)\right|,
\end{aligned}
\tag{17}
$$

where

$$
H_1 = r_L \tau_L (2x_1 + 3) + 3x_1 \quad ; \quad H_2 = -\left(\frac{x_2}{d_C}\right)(x_1(r_L \tau_L + 2) + 2r_L \tau_L + 3)
$$

$$
H_3 = r_L(2x_1 + 3) \quad ; \quad H_4 = -\left(\frac{x_2}{d_C}\right)r_L(x_1 + 2)
\tag{18}
$$

and

$$
x_1 = \left(\frac{\lambda_1}{\lambda_L \tau_1}\right)^2 \frac{r_L\left(\sqrt{3} - \tau_1\right)}{1 + \sqrt{3}\tau_L} \quad ; \quad x_2 = \frac{r_L\left(\sqrt{3} - \tau_1\right)}{\lambda_L^2\left\{\left(1 - \sqrt{3}\tau_L\right) - \tau_1\left(\sqrt{3} + \tau_L\right)\right\}}.
\tag{19}
$$

From (17), it is observed that the series resonances of $|\underline{Y}_{AAk,\text{apx}}|_N$ and $|\underline{Y}_{CCk,\text{apx}}|_N$ admittances match up because their denominators are the same. This is true for the series resonance of $|\underline{Y}_{ACk,\text{apx}}|_N$ admittance and the parallel resonance of $|\underline{Z}_{AAk}|_N$, $|\underline{Z}_{CCk}|_N$ and $|\underline{Z}_{ACk}|_N$ impedances. However, despite the discrepancy in the denominator degrees, it is numerically verified that the harmonic of the parallel and series resonance is roughly the same for all the impedances and admittances. Then, these resonances are located from the minimum value of the $|\underline{Y}_{ACk,\text{apx}}|_N$, $|\underline{Z}_{AAk}|_N$, $|\underline{Z}_{CCk}|_N$ and $|\underline{Z}_{ACk}|_N$ denominator because it is the simplest. In the study, this denominator is labeled as $|\Delta_k|$ for clarity and the harmonic of the parallel and series resonance numerically obtained as the minimum value of $|\Delta_k|$ is labeled as $k_{r,\Delta}$ for both resonances. This value is analytically located by equating to zero the derivative of $|\Delta_k|^2$ with respect to $k$, which can be arranged in the following form:

$$
\frac{\partial\left(\left|\Delta_k\right|^2\right)}{\partial k} = 6H_1 k(k^4 + G_1 k^2 + G_2) \quad \Rightarrow \quad k_{r,a}^4 + G_1 k_{r,a}^2 + G_2 = 0,
\tag{20}
$$

where $k_{r,\,a}$ is the harmonic of the parallel and series resonance analytically obtained and

$$G_1 = \frac{2 \cdot (2H_2H_1 + H_3^2)}{3 \cdot H_1^2} \quad ; \quad G_2 = \frac{H_2^2 + 2 \cdot H_4H_3}{3 \cdot H_1^2}. \tag{21}$$

Thus, the root of equation (20) allows locating the parallel and series resonance:

$$k_{r,a} = \sqrt{\frac{-G_1 + \sqrt{G_1^2 - 4 \cdot G_2}}{2}}. \tag{22}$$

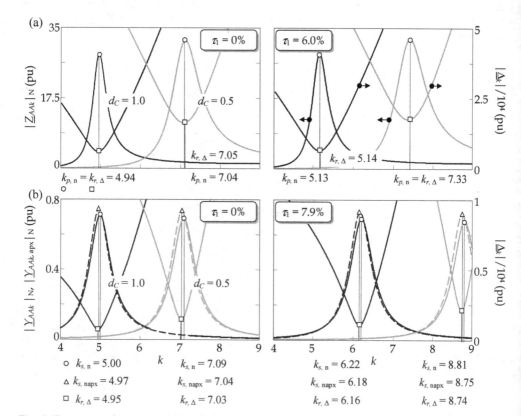

Fig. 8. Resonance location: a) Parallel resonance (power system data in Section 3.1: $X_S$ = 0.049 pu, $R_L$ = 1.341 pu and $\lambda_L$ = 1.0). b) Series resonance (power system data in Section 3.2: $X_S$ = 0.154 pu, $R_L$ = 1.464 pu, $\lambda_L$ = 0.95, $|\underline{Z}_{P1}|$ = 30.788 pu, $\lambda_P$ = 0.95 and three-phase load model LM1).

To illustrate the above study, Fig. 8 shows $|\underline{Z}_{AAk}|_N$, $|\underline{Y}_{AAk}|_N$, $|\underline{Y}_{AAk,\,apx}|_N$ and $|\underline{\Delta}_k|$ for the power systems presented in the laboratory tests of Sections 3.1 and 3.2, and the analytical results of the resonances (22) for these systems are

- Parallel resonance: $k_{r,\,a}$ = 4.94 and 7.05 ($\tau_1$ = 0 and $d_C$ = 1.0 and 0.5, respectively) and $k_{r,\,a}$ = 5.13 and 7.33 ($\tau_1$ = 6.0% and $d_C$ = 1.0 and 0.5, respectively).

- Series resonance: $k_{r,a} = 4.95$ and $7.03$ ($\tau_1 = 0$ and $d_C = 1.0$ and $0.5$, respectively) and $k_{r,a} = 6.16$ and $8.74$ ($\tau_1 = 7.9\%$ and $d_C = 1.0$ and $0.5$, respectively).

From these results, it is seen that

- As the variable $z_P = 30.788/0.154 = 199.9 > 20$, the numerical results obtained from $|\underline{Y}_{AAk,\,apx}|_N$ are similar to those obtained from $|\underline{Y}_{AAk}|_N$, and $k_{s,\,n} \approx k_{s,\,napx}$.
- The harmonic of the $|\underline{Z}_{AAk}|_N$ and $|\underline{Y}_{AAk,\,apx}|_N$ (and therefore $|\underline{Y}_{AAk}|_N$) maximum values nearly coincides with the harmonic of the $|\underline{\Delta}_k|$ minimum value, $k_{r,\,\Delta} \approx k_{p,\,n}$ and $k_{r,\,\Delta} \approx k_{s,\,napx}$, and that (22) provides the harmonic of the parallel and series resonance correctly, i.e. $k_{r,\,a} \approx k_{p,\,n}$ and $k_{r,\,a} \approx k_{s,\,napx}$.
- Although the resistances $R_S$ and $R_1$ of the supply system and the Steinmetz circuit inductor are neglected in the analytical study [i.e., $\underline{Z}_{Sk} \approx j \cdot k \cdot X_S$ and $\underline{Z}_{1k} \approx j \cdot k \cdot X_1$ in (14)], the results are in good agreement with the experimental measurements in Sections 3.1 and 3.2, i.e. $k_{r,\,a} \approx k_{p,\,meas}$ and $k_{r,\,a} \approx k_{s,\,meas}$. However, the magnitude values obtained numerically are greater than the experimental measurements (e.g., $|\underline{Z}_{AAk}| = X_S \cdot |\underline{Z}_{AAk}|_N = 0.049 \cdot 33.03 = 1.62$ pu for $k_{p,\,n} = 7.04$ in the $\tau_1 = 0\%$ and $d_C = 0.5$ plot of Fig. 8 and $|\underline{Z}_{AAk}| \approx 1.1$ pu for $k_{p,\,meas} = 7.2$ in Fig. 5 or $|\underline{Y}_{AAk}| = |\underline{Y}_{AAk}|_N/X_S = 0.73/0.154 = 4.74$ pu for $k_{s,\,n} = 7.09$ in the $\tau_1 = 0\%$ and $d_C = 0.5$ plot of Fig. 8 and $|\underline{Y}_{AAk}| \approx 2.1$ pu for $k_{p,\,meas} = 7.2$ in Fig. 7).
- The influence of the resistance $R_1$ on Steinmetz circuit design (3) shifts the parallel and series resonance to higher frequencies. This was also experimentally verified in the laboratory test of Section 3.

Fig. 9 compares $k_{r,\,a}$, with $k_{p,\,n}$ and $k_{s,\,n}$. Considering the validity range of the involved variables, the values leading to the largest differences are used. It can be observed that $k_{r,\,a}$ provides the correct harmonic of the parallel and series resonance. The largest differences obtained are below 10% and correspond to $k_{s,\,n}$ when $z_P = 20$, which is the lowest acceptable $z_P$ value to apply the $k_{r,\,a}$ analytical expression. Although only the linear load model LM1 is considered in the calculations, it is verified that the above conclusions are true for the other three-phase load models.

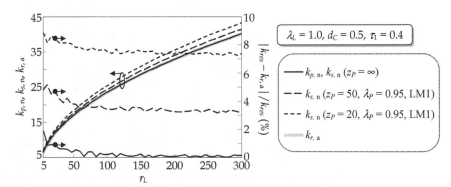

Fig. 9. Comparison between $k_{res}$ and $k_{r,\,a}$ ($k_{res} = k_{p,\,n}$ or $k_{s,\,n}$).

The previous research unifies the study of the parallel and series resonance, providing an expression unique to their location. This expression is the same as in the series resonance case (Sainz et al., 2007), but substantially improves those obtained in the parallel resonance case (Sainz et al., in press). Moreover, the Steinmetz circuit inductor resistance is considered in the analytical location of the resonances, making a contribution to previous studies. This resistance, as well as damping the impedance values, shifts the resonance frequencies because it influences Steinmetz circuit design (i.e., the determination of the Steinmetz circuit reactances).

## 5. Sensitivity analysis of power system harmonic response

A sensitivity analysis of all variables involved in location of the parallel and series resonance is performed from (22). Thus, considering the range of the variables, Fig. 10 shows the contour plots of the harmonics where the parallel and series resonance is located. These harmonics are calculated from the expression of $k_{r,a}$, (22), and the $\tau_1$ range is fixed from (4) considering the $\lambda_L$ value. From Fig. 10, it can be noted that

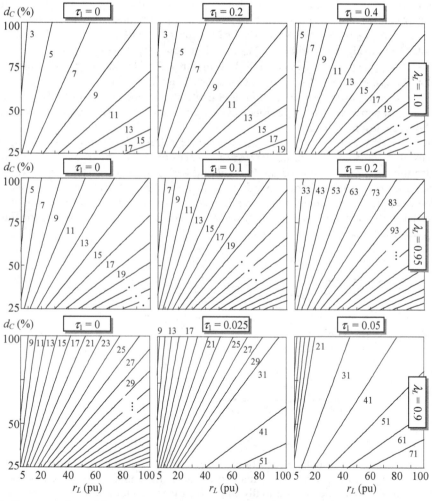

Fig. 10. Contour plots of $k_{r,a}$.

- The resonant behavior of the Steinmetz circuit with power system reactors occurs in a range of relatively large harmonics.
- The resonances are located in the low-order harmonics only if the displacement power factor of the single-phase load impedance is close to the unity value (i.e., $\lambda_L \approx 1$) and this impedance is small in comparison with the supply system reactances (i.e., small $r_L$ ratios). The former condition is common but, considering that $r_L = \lambda_L \cdot S_S / S_L$ (Sainz et al., 2009a), the latter only occurs in weak power systems where the short-circuit power at the PCC bus, $S_S$, is low compared to the apparent power of the single-phase load, $S_L$.
- The resonances are shifted to high-order harmonics if the $\tau_1$ ratio of the Steinmetz circuit inductor is far from the zero value, i.e. its displacement power factor $\lambda_1$ is far from the unity value. It is also true if the Steinmetz circuit capacitor degrades, i.e. the Steinmetz circuit suffers capacitor loss and $d_C$ is also far from the unity value.

# 6. Examples

For the sake of illustration, two different implementations of the $k_{r, a}$ expression, (22), are developed. In the first, the analytical study in Section 4 is validated from laboratory measurements. Several experimental tests were made to check the usefulness of the $k_{r, a}$ expression in locating the parallel and series resonance. In the second, this expression is applied to locate the harmonic resonance of several power systems with a Steinmetz circuit in the literature.

## 6.1 Experimental measurements of power system harmonic response

To validate the analytical study, measurements were made in two downscaled laboratory systems corresponding to the networks of Fig. 4 (parallel resonance) and Fig. 6 (series resonance). The frequency response measurements were made with a 4.5 kVA AC ELGAR Smartwave Switching Amplifier as the power source, which can generate sinusoidal waveforms of arbitrary frequencies (between 40 Hz and 5000 Hz) and a YOKOGAWA DL 708 E digital scope as the measurement device. From the results shown in the next Sections, it must be noted that (22) provides acceptable results. Although experimental tests considering the inductor resistance ($R_1 \approx 0.1342$ pu) are not shown, they provide similar results.

## 6.1.1 Experimental measurements of the parallel resonance

The harmonic response of the network in Fig. 4 was measured in the laboratory for two cases with the following system data ($U_B = 100$ V and $S_B = 500$ VA):

- Case 1 (studied in Section 3.1):
  - Supply system: $\underline{Z}_{S1} = 0.022 + j0.049$ pu.
  - Railroad substation: $R_L = 1.341$ pu, $\lambda_L = 1.0$.
  - External balancing equipment: $X_{1, \text{apx}} = 2.323$ pu and $X_{2, \text{apx}} = 2.323$ pu [neglecting the inductor resistance, (1)] and $d_C = 1.0, 0.75, 0.5$ and $0.25$.
- Case 2: System data of Case 1 except the single-phase load fundamental displacement factor of the railroad substation, which becomes $\lambda_L = 0.95$. The Steinmetz circuit reactances also change, i.e. $X_{1, \text{apx}} = 1.640$ pu and $X_{2, \text{apx}} = 5.975$ pu (1).

Fig. 11a compares the parallel resonance measured in the experimental tests with those obtained from (22). In order to analytically characterize the resonance, the variable values corresponding to the above data are $r_L = 27.4$, $\lambda_L = 1$ and $0.95$ (Cases 1 and 2, respectively) and $\tau_1 = 0$.

### 6.1.2 Experimental measurements of the series resonance

The harmonic response of the network in Fig. 6 was measured in the laboratory for two cases with the following system data ($U_B = 100$ V and $S_B = 500$ VA):

- Case 1:
  - Supply system: $\underline{Z}_{S1} = 0.076 + j0.154$ pu.
  - Railroad substation: $R_L = 1.464$ pu, $\lambda_L = 1.0$.
  - External balancing equipment: $X_{1,\,apr} = 2.536$ pu and $X_{2,\,apr} = 2.536$ pu [neglecting the inductor resistance, (1)] and $d_C = 1.0$, 0.75, 0.5 and 0.25.
  - Three-phase load: Grounded wye series R-L impedances with $|\underline{Z}_{P1}| = 30.788$ pu and $\lambda_P = 0.95$ are connected, i.e. the three-phase load model LM1 in (Task force on Harmonic Modeling and Simulation, 2003).
- Case 2 (studied in Section 3.2): System data of Case 1 except the single-phase load fundamental displacement factor of the railroad substation, which becomes $\lambda_L = 0.95$. The Steinmetz circuit reactances also change, i.e. $X_{1,\,apr} = 1.790$ pu and $X_{2,\,apr} = 6.523$ pu (1).

Fig. 11b compares the series resonance measured in the experimental tests with those obtained from (22). In order to analytically characterize the resonance, the variable values corresponding to these data are $r_L = 9.51$, $\lambda_L = 1$ and $0.95$ (Cases 1 and 2, respectively) and $\tau_1 = 0$.

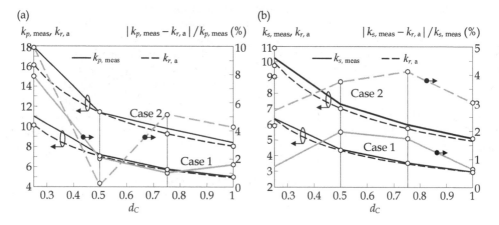

Fig. 11. Comparison between $k_{res}$ and $k_{r,\,a}$. a) $k_{res} = k_{p,\,meas}$. b) $k_{res} = k_{s,\,meas}$.

### 6.2 Harmonic resonance location in several power systems

This section briefly describes several works in the literature on the Steinmetz circuit in power systems, and determines the harmonic of the resonance produced by the presence of this circuit from (22). This allows interpreting the results in the works and predicting the harmonic behavior of the studied power systems.

In (ABB Power Transmission, n.d.), an extensive railway network for coal haulage in East Central Queensland is presented and the installation of nine SVCs in the 132 kV grid to achieve dynamic load balancing is analyzed. The traction load is supplied from single-phase 132/50 kV transformers at each supply substation providing a 25 kV catenary voltage from 50/25 kV autotransformers at intervals along the track. The short-circuit power $S_S$ at 132 kV bus is below 300 MVA while traction loads may reach short duration peaks of $S_L$ = 20 to 40 MVA. A total of 28 single-phase harmonic filters for 50 kV tuned to the 3rd, 5th and 7th harmonics were installed in the substations to prevent harmonics generated in the locomotive thyristor drives from being injected into the 132 kV power system. The harmonic impact of the Steinmetz circuit installation on this traction system can be examined from (22). Considering $\tau_1$ = 0, $d_C$ = 1 and the displacement power factor $\lambda_L$ of the traction load close to the unity value, the ratio $r_L = R_L/X_S = \lambda_L \cdot S_S/S_L$ is between 15 to 7.5 ($S_L$ = 20 to 40, respectively) and the resonance is located at the harmonics $k_{r,a}$ = 3.7 to 2.68. It is interesting to note that the Steinmetz circuit connection could cause parallel and series resonances close to the 3rd harmonic, damaging harmonic power quality. If the displacement power factor was below unity value (e.g., $\lambda_L$ = 0.95), the resonance would shift to $k_{r,a}$ = 5.93 to 4.36 ($S_L$ = 20 to 40, respectively) worsening the harmonic problem. In conclusion, it is not advisable to use the Steinmetz circuit to balance the traction load currents consumed in this installation. However, since the short-circuit power can be below 300 MVA and the transformer short-circuit impedances are not considered in the study, the ratio $r_L$ values can be lower than the previous ones and the resonance can be below the 3rd harmonic (see Fig. 10) avoiding harmonic problems.

In (Barnes & Wong, 1991), an unbalance and harmonic study carried out for the Channel Tunnel 25 kV railway system supplied from the UK and French 400/225/132 kV grid systems is presented. On the UK side, the PCC between the traction load and the tunnel auxiliary load is at the Folkestone 132 kV busbar with a minimum short-circuit power $S_S$ equal to 800 MVA. On the French side, the PCC between the traction load, the auxiliary load and other consumers is at the Mandarins 400 kV busbar with a minimum short-circuit power $S_S$ equal to 11700 MVA. The traction loads range from $S_L$ = 0 to 75 MVA with a displacement power factor $\lambda_L$ = 0.93. Steinmetz circuit is located on the UK side with fast-acting thyristor-controlled reactors and capacitors, which enable the balancing equipment output to vary with the load pattern. Moreover, harmonic studies based on the harmonic spectrum measured in the catenaries of the British Rail network and provided by continental locomotive manufacturers were conducted to analyze the harmonic filter installation. They revealed that the harmonic limits on the French side are within specification limits and no filters are required while, on the UK side, these limits are exceeded and harmonic filters must be installed to reduce harmonic distortion to acceptable levels. These studies can be complemented with harmonic resonance location in the Steinmetz circuit. Thus, considering $\tau_1$ = 0, $d_C$ = 1 and the maximum traction load (i.e., $S_L$ = 75 MVA), the ratio $r_L = R_L/X_S = \lambda_L \cdot S_S/S_L$ is 145.08 and 9.92 and the resonance is located at harmonics $k_{r,a}$ = 21.6 and 6.0 on the French and UK side, respectively. This resonance is shifted to higher harmonics if the traction load is lower. The auxiliary loads and other consumers are not considered in the location of the resonances because their impedance is large enough (i.e., $z_P$ > 20).

In (Arendse & Atkinson-Hope, 2010), the design of the Steinmetz circuit in unbalanced and distorted power supplies is studied from a downscaled laboratory system such as that in Fig. 3. The system data are $\underline{Z}_{S1}$ = 0.0087 +$j$0.00079 $\Omega$, $R_L$ = 4.84 $\Omega$, $\lambda_L$ = 1.0, $\tau_1$ = 0, $d_C$ = 1.0 and a three-phase Variable Speed Drive (VSD) of 24 kVA rated power is used as a harmonic source. A three-phase linear load with $|\underline{Z}_{P1}|$ = 9.802 Ω and $\lambda_P$ = 0.81 (load model LM1) is also connected. The study shows that there is no harmonic problem in the system and that voltage distortion is below 0.05% [Table 7 in (Arendse & Atkinson-Hope, 2010)]. This can be analyzed from (22) because, considering that $r_L$ = 4.84/0.00079 = 6127 and $z_P$ = 9.802/0.00079 = 12408 (i.e., the three-phase linear load influence is negligible), the parallel resonance "observed" from the VSD is located at $k_{r,a}$ = 72.9.

## 7. Conclusion

In this chapter, the analytical study conducted in previous works on the parallel and series resonance in power systems with a Steinmetz circuit is unified and an expression unique to the location of both resonances is provided, which substantially improves those proposed in earlier works on the parallel resonance. This expression considers not only the impact of capacitor degradation on the resonance but also the resistance of the Steinmetz circuit inductor, which is another contribution to previous studies.

The sensitivity analysis reveals that the resonances mainly depend on the power system inductors and the single-phase load of the Steinmetz circuit. However, capacitor bank degradation and the R/X ratio of the Steinmetz circuit inductor can also strongly influence the resonance. Broadly speaking, Steinmtez circuit resonances with power system reactors appear at high-order harmonics. They only occur at low-order harmonics if the single-phase load impedance is small in comparison with the supply system reactance (i.e., in weak power systems) and the single-phase load power displacement factor is close to the unity value. The study also shows that the capacitor bank degradation and the resistance of the Steinmetz circuit inductor shift the resonance to higher harmonics. The analytical study results are validated with experimental measurements in a downscaled laboratory system and the study is applied to analyze several power systems with a Steinmetz circuit in the literature. Measurements in actual ac traction systems will be necessary to fully confirm these results.

Future research should focus on the power system harmonic response "observed" from the railroad substation. The framework developed in the previous research and completed in this Chapter must make it possible to obtain analytical expressions to locate resonances from the substation.

## 8. Acknowledgment

This work is supported by grant DPI2010-15448.

## 9. References

ABB Power Transmission (n.d.). Multiple SVC installations for traction load balancing in Central Queensland. In: *Pamphlet A02-0134*, 26/02/2011, Available from <http://www.abb.com/>.

Arendse, C. & Atkinson-Hope, G. (2010). Design of a Steinmetz symmetrizer and application in unbalanced network. *Proceedings of the 45th International Universities Power Engineering Conference (UPEC)*, pp. 1-6, 2010.

Barnes, R. & Wong, K. T. (1991). Unbalance and harmonic studies for the Channel Tunnel railway system. *IEE Proceedings B, Electric Power Applications*, Vol. 138, No. 2, 1991, pp. 41-50.

Capasso, A. (1998). The power quality concern in railway electrification studies. *Proceedings of 8th IEEE Int. Conf. on Harmonics and Quality of Power (ICHQP)*, pp. 647-652, 1998.

Caro, M., Sainz, L. & Pedra, J. (2006). Study of the power system harmonic response in the presence of the Steinmetz circuit. *Electric Power Systems Research*, Vol. 76, No. 12, August 2006, pp. 1055-1063.

Chen, T-H. (1994). Criteria to estimate the voltage unbalances due to high-speed railway demands. *IEEE Transactions on Power Systems*, Vol. 9, No. 3, August 1994, pp. 1672-1678.

Chen, T-H. & Kuo, H-Y. (1995). Analysis on the voltage unbalance due to high-speed railway demands. *Proceedings of the International Conference on Energy Managment and Power Delivery*, pp. 657-661, 1995.

Chicco, G., Chindris, M., Cziker, A., Postolache, P. & Toader, C. (2009). Analysis of the Steinmetz compensation circuit with distorted waveforms through symmetrical component-based indicators. *Proceedings of the IEEE Bucharest Power Tech Conference 2009*, pp. 1-6, 2009.

Chindris, M., Cziker, A., Stefanescu, A. S. & Sainz, L. (2002). Fuzzy logic controller for Steinmetz circuitry with variable reactive elements. *Proceedings of 8th International Conference OPTIM 2002*, Proc. 1G.3, pp. 233-238, 2002.

Czarnecki, L. S. (1989). Reactive and unbalanced currents compensation in three-phase asymmetrical circuits under non-sinusoidal conditions. *IEEE Transactions on Instrumentation and measurement*, June 1989, Vol. 38, No. 3, pp. 754-759.

Czarnecki, L. S. (1992). Minimization of unbalanced and reactive currents in three-phase asymmetrical circuits with non-sinusoidal voltage. *Proceedings IEE*, Vol. 139, Pt. B., No. 4, July 1992, pp. 347-354.

Hill, R. J. (1994). Electric railway traction. Part3: Traction power supplies. *Power Engineering Journal*, Vol. 8, No. 6, 1994, pp. 275-286.

Howroyd, D. C. (1989). Public supply disturbances from AC traction. *Proceedings of the International Conference on Main Line Railway Electrification*, pp. 260-264, 1989.

IEC 61000-3-6, Part 3-6: Limits – Assessment of emission limits for the connection of distorting installations to MV, HV and EHV power systems, 2008-02.

Jordi, O., Sainz, L. & Chindris, M. (2002). Steinmetz system design under unbalanced conditions. *European Transactions on Electrical Power*, Vol. 12, No. 4, July/August 2002, pp. 283-290.

Lee, S.Y. & Wu, C.J. (1993). On-line reactive power compensation schemes for unbalanced three-phase four wire distribution feeders. *IEEE Transactions on Power Delivery*, Vol. 8, No. 4, October 1993, pp. 1958-1965.

Marczewski, J. J. (1999). IEEE working group on system and equipment considerations for traction. Utility interconnection issues. *Proceedings of IEEE Power Engineering Society Summer Meeting*, Vol. 1, pp. 439-444, 1999.

Mayer, D. Kropik, P. (2005). New approach to symmetrization of three-phase networks. *Journal of Electrical Engineering*, 2005, Vol. 56, No. 5-6, pp. 156-161.

Qingzhu, W., Mingli, W., Jianye, C. & Guipping, Z. (2010). Optimal balancing of large single-phase traction load. *Proceedings of the IET Conference on Railway Traction Systems (RTS 2010)*, pp. 1-6, 2010.

Qingzhu, W., Mingli, W., Jianye, C. & Guipping, Z. (2010). Model for optimal balancing single-phase traction load based on the Steinmetz's method. *Proceedings of the IEEE Energy Conversion Congress an Exposition (ECCE)*, pp. 1565-1569, 2010.

Sainz, L., Caro, M. & Pedra, J. (2004). Study of electric system harmonic response. *IEEE Transactions on Power Delivery*, Vol. 19, No. 2, April 2004, pp. 868-874.

Sainz, L., Pedra, J. & Caro, M. (2005). Steinmetz circuit influence on the electric system harmonic response. *IEEE Transactions on Power Delivery*, Vol. 20, No. 2, April 2005, pp. 1143-1156.

Sainz, L., Pedra, J. & Caro, M. (2007). Influence of the Steinmetz circuit capacitor failure on the electric system harmonic response. *IEEE Transactions on Power Delivery*, Vol. 22, No. 2, April 2007, pp. 960-967.

Sainz, L., Pedra, J. & Caro, M. (2009). Background voltage distortion influence on the power electric systems in the presence of the Steinmetz circuit. *Electric Power Systems Research*, Vol. 79, No. 1, January 2009, pp. 161-169.

Sainz, L., Caro, M. & Caro, E. (2009). Analytical study on the series resonance in power systems with the Steinmetz circuit. *IEEE Transactions on Power Delivery*, Vol. 24, No. 4, October 2009, pp. 2090-2098.

Sainz, L., Caro, M., Caro, E. (in press). Influence of Steinmetz Circuit Capacitor Degradation on Series Resonance of Networks. *European Transactions on Electrical Power*, in press (DOI: 10.1002/etep.514).

Sainz, L. & Riera, S. (submitted for publication). Study of the Steinmetz circuit design. *Power Systems Research*.

Task Force on Harmonics Modeling and Simulations. Modeling and simulation of the propagation of harmonics in electric power networks. Part I: Concepts, models and simulation techniques. *IEEE Transactions on Power Delivery*, Vol. 11, No. 1, January 1996, pp. 452–465.

Task Force on Harmonic Modeling and Simulation. Impact of aggregate linear load modeling on harmonic analysis: A comparison of common practice and analytical models. *IEEE Transactions on Power Delivery*, Vol. 18, No. 2, April 2003, pp. 625-630.

# Part 4

## Industrial Environments

# 9

# Harmonic Distortion in Renewable Energy Systems: Capacitive Couplings

Miguel García-Gracia, Nabil El Halabi,
Adrián Alonso and M.Paz Comech
*CIRCE (Centre of Research for Energy Resources and Consumption)*
*University of Zaragoza*
*Spain*

## 1. Introduction

Renewable energy systems such as wind farms and solar photovoltaic (PV) installations are being considered as a promising generation sources to cover the continuous augment demand of energy.

With the incoming high penetration of distributed generation (DG), both electric utilities and end users of electric power are becoming increasingly concerned about the quality of electric network (Dugan et al., 2002). This latter issue is an umbrella concept for a multitude of individual types of power system disturbances. A particular issue that falls under this umbrella is the capacitive coupling with grounding systems, which become significant because of the high-frequency current imposed by power converters.

The major reasons for being concerned about capacitive couplings are:

a. Increase the harmonics and, thus, power (converters) losses in both utility and customer equipment.
b. Ground capacitive currents may cause malfunctioning of sensitive load and control devices.
c. The circulation of capacitive currents through power equipments can provoke a reduction of their lifetime and limits the power capability.
d. Ground potential rise due to capacitive ground currents can represent unsafe conditions for working along the installation or electric network.
e. Electromagnetic interference in communication systems and metering infrastructure.

For these reasons, it has been noticed the importance of modelling renewable energy installations considering capacitive coupling with the grounding system and thereby accurately simulate the DC and AC components of the current waveform measured in the electric network.

Introducing DG systems in modern distribution networks may magnify the problem of ground capacitive couplings. This is because DG is interfaced with the electric network via power electronic devices such as inverters.

These capacitive couplings are part of the electric circuit consisting of the wind generator, PV arrays, AC filter elements and the grid impedance, and its effect is being appreciated in most large scale DG plants along the electric network (García-Gracia et al., 2010).

Power electronic devices, as used for DG, might be able to cause harmonics. The magnitude and the order of harmonic currents injected by DC/AC converters depend on the technology of the converter and mode of its operation (IEC Std. 61000-4-7, 2010, IEEE Std. 519-1992, 1992).

Due to capacitive coupling between the installation and earth, potential differences imposed by switching actions of the converter inject a capacitive ground current which can cause significant electromagnetic interferences, grid current distortion, losses in the system, high-noise level in the installation and unsafe work conditions (Chicco et al., 2009).

Several renewable system installations analyses have been reported (Bellini, 2009, Conroy, 2009, Luna, 2011, Sukamonkol, 2002, Villalva, 2009), where most theoretical analysis and experimental verifications have been performed for small-scale installations without considering capacitive coupling. Power electronics models and topologies also have been studied, but without considering the amount of losses produced by the capacitive current that appears due to the switching actions (Zhow, 2010, Chayawatto, 2009, Kim, 2009). In (Iliceto & Vigotti, 1998), the total conversion losses of a real 3 MW PV installation have been studied considering reflection losses, low radiation and shadow losses, temperature losses, auxiliary losses, array losses and converters losses. The latter two factors sum a total of 10% of the rated power where part of these losses is due to the capacitive coupling that was neglected.

Therefore, for an accurate study of power quality, it is important to model DG installations detailing the capacitive coupling of the electric circuit with the grounding system, which are detailed for PV installations and wind farms in Sections 2 and 3, respectively. These models allow analyzing the current distortion, ground losses and Ground Potential Rise (GPR) due to the capacitive coupling. The combined effect of several distributed generation sources connected to the same electric network has been simulated, and results have been presented together with solutions based on the proposed model to minimize the capacitive ground current for meeting typical power quality regulations concerning to the harmonic distortion and safety conditions.

## 2. Capacitive coupling in solar-photovoltaic installation

The region between PV modules and PV structure essentially acts as an insulator between layers of PV charge and ground. Most shunt capacitive effects that may be ignored at very low frequencies can not be neglected at high frequencies for which the reactance will become relatively small due to the inverse proportionality with frequency $f$ and, therefore, a low impedance path is introduced between power elements and ground.

This effect is present in PV installations because of the high frequency switching carried out by the converters stage, which arises different capacitive coupling between modules and ground. Thus, the capacitive effect must be represented as a leakage loop between PV arrays, cables and electronic devices and the grounding system. By means of this leakage loop, capacitive currents are injected into the grounding system creating a GPR along the PV installation which introduces current distortion, electromagnetic interference, noise and unsafe work conditions. For this reason, an accurate model of these capacitive couplings are requiered for PV installations.

### 2.1 Equivalent electric circuit for ground current analysis
Depending on the switching frequency, the harmonics produced may be significant according to the capacitive coupling and the resonant frequency inside the PV installation.

Moreover, every PV array is considered as an independent current source with a DC current ripple independent of the converter ripple. These ripple currents are not in synchronism with the converter and produce subharmonics in the DC circuit which increase the Total Harmonic Distortion in the current waveform (THDI) (Zhow et al., 2010).

The typical maximum harmonic order h = 40, defined in the power quality standards, corresponds to a maximum frequency of 2 kHz (with 50 Hz as fundamental frequency) (IEC Std. 61000-4-7, 2002). However, the typical switching frequency of DC/DC and DC/AC converters, usually operated with the Pulse Width Modulation (PWM) technique, is higher than 3 kHz. Hence, higher order harmonics up to the 100th order, can be an important concern in large scale PV installations where converters with voltage notching, high pulse numbers, or PWM controls result in induced noise interference, current distortion, and local GPR at PV arrays (Chicco et al., 2009).

A suitable model of capacitive couplings allows reproducing these harmonic currents injected not only into the grid, but also into the DC circuit of the PV installation that would lead to internal resonant, current distortion and unsafe work conditions where capacitive discharge currents could exceed the threshold of safety values of work (IEEE Std. 80-2000, 2000). The capacitive coupling is part of the electric circuit consisting of the PV cells, cables capacitive couplings, AC filter elements and the grid impedance, as shown in Fig. 1, and its effect is being appreciated in most large scale PV plants.

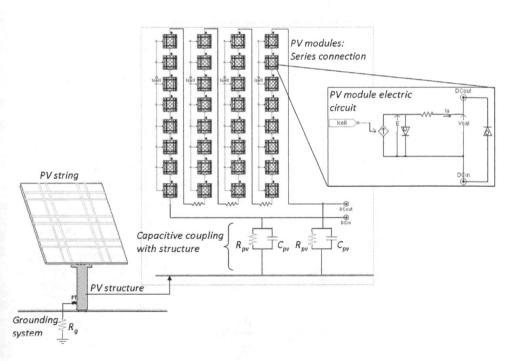

Fig. 1. Model of PV module, PV array and capacitive coupling with PV structure.

## 2.2 Behavior of the PV installation considering capacitive coupling

Normally, numerous PV modules are connected in series on a panel to form a PV array as it is shown in Fig. 1. The circuit model of the PV module (Kim et al., 2009) is composed of an ideal current source, a diode connected in parallel with the current source and a series resistor. The output current of each PV module is determined as follows:

$$I = I_{sc} - I_d = I_{sc} - I_o \cdot \left[ \exp\left( \frac{V + I \cdot R_s}{n \cdot V_T} \right) \right] \tag{1}$$

where $I_o$ is the diode saturation current, $V$ the terminal voltage of a module, n the ideal constant of diode, $V_T$ is the thermal potential of a module and it is given by $m \cdot (kT/q)$ where $k$ the Boltzmann's constant (1.38E-23 J/K), $T$ the cell temperature measured in K, $q$ the Coulomb constant (1.6E-19 C), and $m$ the number of cells in series in a module. $I_{sc}$ is the short circuit current of a module under a given solar irradiance. $I_d$ is the diode current, which can be given by the classical diode current expression. The series resistance $R_s$ represents the intrinsic resistance to the current flow.

The capacitive coupling of PV modules with the ground is modelled as a parallel resistance $R_{pv}$ and capacitor $C_{pv}$ arrangement which simulates the frequency dependency on the insulator between PV modules and the grounding system. The PV structure is connected to the grounding system represented in the model by the grounding resistance $R_g$.

Taking into account that the converter represents a current source for both DC circuit and AC circuit of the PV installation, an equivalent circuit is deduced to analyze the capacitive coupling effect over the current and voltage waveforms.

The equivalent circuit of both DC circuit of the PV installation and AC circuit for connection to the grid as seen between inverter terminals and ground is illustrated Fig. 2. In the AC circuit $R_{ac\_cable}$, $L_{ac\_cable}$ and $C_{ac\_cable}$ are the resistance, inductance and capacitance of the AC underground cables, $R_{g\_es}$ is the ground resistance at the substation and $L_{filter}$ and $C_{filter}$ are the parameters of the LC filter connected at AC terminals of the inverter.

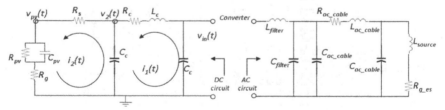

Fig. 2. Capacitive coupling model for the DC and AC electric circuit of a PV installation.

The inclusion of $R_{pv}$ and $C_{pv}$ on the PV equivalent circuit allows representing the leakage path for high frequency components between PV modules and ground. This DC equivalent circuit is represented by the following continuous-time equations, at nominal operating condition

$$\frac{di_1(t)}{dt} = \frac{1}{L_c} \cdot v_{in}(t) - \frac{R_c}{L_c} \cdot i_1(t) - \frac{1}{L_c} \cdot v_2(t) \tag{2}$$

$$\frac{di_2(t)}{dt} = \frac{1}{C_c \cdot (R_s + R_g)} \cdot i_1(t) - \left[ \frac{1}{C_c \cdot (R_s + R_g)} + \frac{\psi}{C_{pv} \cdot R_{pv}} \right] \cdot i_2(t) + \frac{1}{C_{pv} \cdot R_{pv} \cdot (R_s + R_g)} \cdot v_2(t) \tag{3}$$

$$\frac{dv_2(t)}{dt} = \frac{1}{C_c} \cdot i_1(t) - \frac{1}{C_c} \cdot i_2(t) \tag{4}$$

$$\frac{dv_{pv}(t)}{dt} = \frac{R_g}{C_c \cdot (R_s + R_g)} \cdot i_1(t) + \left( \frac{R_{pv} + R_g}{C_{pv} \cdot R_{pv}} - \frac{R_g}{C_c \cdot (R_s + R_g)} - \frac{R_g \cdot \psi}{C_{pv} \cdot R_{pv}} \right) \cdot i_2(t) \ldots$$
$$+ \frac{R_g}{C_{pv} \cdot R_{pv} \cdot (R_s + R_g)} \cdot v_2(t) - \frac{1}{C_{pv} \cdot R_{pv}} \cdot v_{pv}(t) \tag{5}$$

According to the equivalent DC circuit shown in Fig. 2, $i_1(t)$ and $i_2(t)$ are the current of mesh 1 and mesh 2, respectively, $v_{in}(t)$ is the injected voltage by the converter, $v_2(t)$ the voltage at node 2 and $v_{pv}(t)$ is the voltage between PV module and ground and represents the parameter under study. Parameter $C_c$ represents the capacitive coupling between cables and ground and $R_c$ and $L_c$ are the resistance and inductance of the cable, respectively.

In some simplified models of PV installation (Villalva, 2009, Kim, 2009, Bellini, 2009), the capacitance $C_{pv}$ and resistance $R_{pv}$ are considered like infinite and zero, respectively. Then, the capacitive coupling with the grounding system is totally neglected. Even ground resistance of the PV installation is not considered ($R_g = 0$).

The PV arrays are connected to a DC system of 700 V, and the power is delivered to an inverter stage based on four inverters of 125 kW in full bridge topology and operation frequency of 3.70 kHz. The underground cables, which connect the PV arrays to the inverter stage, have been included through their frequency dependent model.

The electrical parameters of the capacitive coupling between PV arrays and grounding system are shown in Table 1 and have been adjusted according to the field measurement in order to simulate the response of the capacitive coupling model accurately against the harmonics injected by the operation of the converters.

| Element | Parameter | Value |
|---|---|---|
| PV array | Operation voltage | 700 Vdc |
| | Capacitance $C_{PV}$ | $1 \times 10^{-9}$ F |
| | Resistance $R_{PV}$ | $1 \times 10^7$ $\Omega$ |
| | Series Resistance $R_S$ | 0.30 $\Omega$ |
| DC cable | Resistance $R_C$ | 0.25 $\Omega$/km |
| | Inductance $L_C$ | 0.00015 H/km |
| | Capacitive coupling $C_C$ | $1 \times 10^{-4}$ F/km |
| Filter LC | Inductance $L_{filter}$ | 90 $\mu$H |
| | Capacitance $C_{filter}$ | 0.756 mF |
| Underground cable | Positive sequence impedance | $0.3027 + j\,0.1689$ $\Omega$/km |
| | Zero sequence impedance | $0.4503 + j\,0.019$ $\Omega$/km |
| | Zero sequence susceptance | 0.1596 mS/km |
| Power grid | Thevenin voltage | 20 kV |
| | Thevenin impedance | $0.6018 + j\,2.4156$ $\Omega$ |
| | Ground resistance $R_g$ | 1.2 $\Omega$ |

Table 1. Electric parameters for the solar PV installation capacitive grounding model.

The frequency response of both capacitive coupling and simplified model for the DC circuit of a PV installation operating at nominal operating condition is shown in the Bode diagram of Fig. 3. The capacitive coupling model presents a considerable gain for waveforms under 108 kHz in comparison with the simplified model which has a limited gain for this range of frequencies. Hence, the capacitive coupling model is able to simulate the leakage loop between PV module and grounding system for high frequencies, unlike the simplified model.

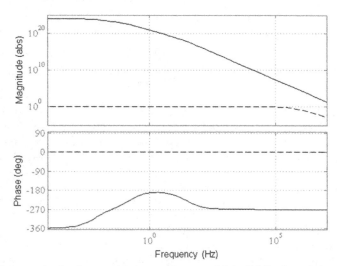

Fig. 3. Bode diagram for both capacitive coupling model (solid line) and simplified model (dashed line) for the DC circuit of a PV installation.

The PV installation modelled consists of 184 PV arrays connected in parallel to generate 1 MW. Both circuits have been modelled to analyze the mutual effect raised from the capacitive couplings between the electric circuits and the grounding system, at rated operating conditions. The simulation has been performed using PSCAD/EMTDC [PSCAD, 2006] with a sampling time of at least 20 μs.

The current and voltage waveforms obtained from the proposed model together with the FFT analysis are shown in Fig. 4a and Fig. 4b. The THDI and THDV obtained from simulations are 26.99% and 3725.17%, respectively, where the DC fundamental component of current is 88.56 mA, and the fundamental voltage component is 8.59 V. The frequencies where most considerable harmonic magnitudes are the same of those obtained at field measurement; 3.70 kHz, 11.10 kHz, 14.80 kHz and 18.50 kHz within a percentage error of ±27.42% for fundamental component and ±15.35% for the rest of harmonic components.

## 2.3 Additional information provided by the PV installation capacitive coupling

The model considering capacitive coupling between PV modules and grounding system of the installation leads to an accurate approximation to the response of the PV installation against the frequency spectrum imposed by the switching action of the inverters. This approximation is not feasible using simplified models because of the bandwidth limitation shown in Fig. 3 for high frequencies.

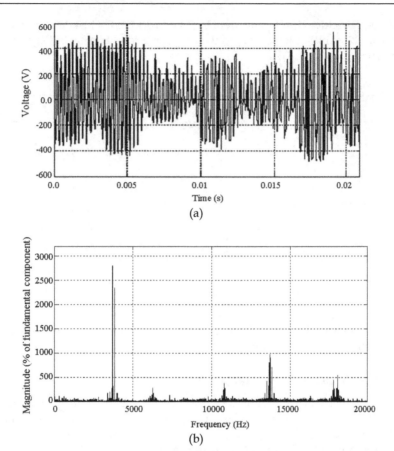

Fig. 4. Simulation result of the capacitive coupling model: (a) voltage waveform between PV array and grounding system and (b) FFT analysis of the voltage waveform obtained.

Simulation results indicate that ground current in large scale PV installations can be considerable according to the values expressed in (IEEE Std. 80-2000, 2000). In the range of 9-25 mA range, currents may be painful at 50-60 Hz, but at 3-10 kHz are negligible (IEC 60479-2, 1987). Thus, the model allows the detection of capacitive discharge currents that exceeds the threshold of safety values at work.

Because of large scale installations are systems with long cables, the resonant frequency becomes an important factor to consider when designing the AC filters and converters operation frequency. The proposed model accurately detects the expected resonant frequency of the PV installations at 12.0 kHz with an impedance magnitude Z of 323.33 Ω while simplified models determine a less severe resonant at a frequency value of 15.50 kHz with a Z of 150.45 Ω, as shown in Fig. 5.

This latter resonant frequency is misleading and pointless for the real operating parameters of the installation. The total DC/AC conversion losses obtained from simulations is 5.6% when operating at rated power, which is equivalent to 56.00 kW. Through the proposed model, it has been detected that a 22.32% of the losses due to the DC/AC conversion is because of the capacitive coupling modelled. Thus, a 1 MW PV installation as modelled in Fig. 2 presents 12.50 kW of losses due to the capacitive couplings or leakage loop between PV modules and ground.

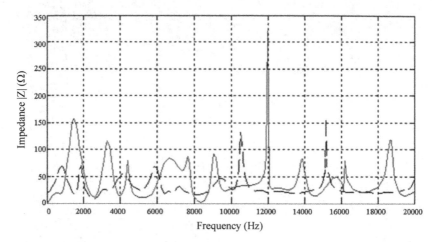

Fig. 5. Resonance frequency of the PV installation without capacitive coupling (dashed line) and considering capacitive couplings (solid line).

## 3. Capacitive coupling in wind farms

Wind energy systems may contribute to the distribution network voltage distortion because of its rotating machine characteristics and the design of its power electronic interface. As presented in Fig. 6, wind energy system designs incorporate a wide range of power electronic interfaces with different ratings (Comech et al., 2010).

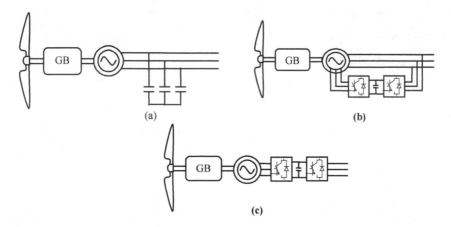

Fig. 6. Wind turbine configurations.

Fig. 6a shows the fixed-speed wind turbine with asynchronous squirrel cage induction generator (SCIG) directly connected to the grid via transformer. Fig. 6b represents the limited variable speed wind turbine with a wound rotor induction generator and partial scale frequency converter on the rotor circuit known as doubly fed induction generator (DFIG). Fig. 6c shows the full variable speed wind turbine, with the generator connected to the grid through a full-scale frequency converter.

These power electronic interfaces are rated as a percentage of the machine power, hence larger systems are accountable for higher distortions. Recent investigations based on wind energy systems suggests that frequency converters (with a typical pulse width modulated with 2.5 kHz of switching frequency) can, in fact, cause harmonics in the line current, leading to harmonic voltages in the network (Conroy & Watson, 2009).

Moreover, most simplified models of wind farms consider a simple series impedance model for underground cables that connect wind turbines with the network grid. Thus, capacitive couplings with ground through cables are not considered for different frequencies components.

To simulate wind farms harmonic distortion behaviour accurately, it is important to model cables by their frequency dependent model. The equivalent circuit for the capacitive coupling model of wind farms is shown in Fig. 7.

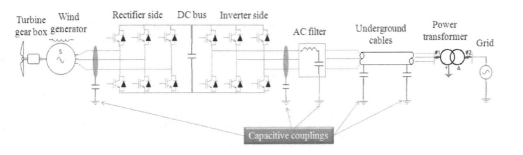

Fig. 7. Capacitive coupling model for wind farm.

Notice that, otherwise the capacitive model of solar installations, the wind turbine is directly connected to the rectifier side of the converter. The capacitive coupling seen by the DC bus through the wind turbine is composed of the path between the rectifier side and ground because of the high harmonic current component imposed by the switching actions, whereas the capacitive coupling seen through the grid is represented by the inverter side, the filter and the underground cable. The equivalent electric circuit of the wind farm capacitive coupling model is shown in Fig. 8.

In this figure, parameters $R_{WG}$ and $L_{WG}$ make reference to the resistance and inductance, respectively, of the synchronous wind generator. $R_g$ is the ground resistance at the wind turbine location while $R_{q\_es}$ is the ground resistance of the electrical substation belonging to the wind farm under study.

$C_{rectifier}$ and $C_{ac\_cable}$ are the capacitive couplings of the rectifier side and underground cable, respectively, with ground. $R_{ac\_cable}$ and $L_{ac\_cable}$ make reference to the resistance and inductance, respectively, of the synchronous wind generator. $L_{filter}$ and $C_{filter}$ are the dimensions of the filter. $L_{TR}$ is the equivalent impedance of the power transformer and $L_{source}$ the thevenin impedance of the source. The variables $v_{WT}(t)$ and $v_{source}(t)$ are the voltages at wind generator node and network grid source, respectively. The input voltage $v_{in}(t)$ is the voltage injected into the grid by the inverter side.

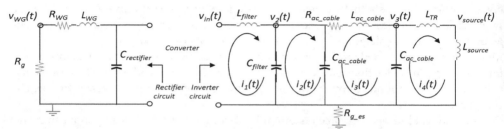

Fig. 8. Equivalent electric circuit belonging to the wind farm capacitive coupling.

The state variable representing this model can be deduced in a similar way as expressed in Section 2. Nonetheless, the effect of capacitive couplings in wind farms is more significant at the inverter circuit through the power grid where the circuit of the filters and cables exert an important influence over the ground currents.

The continuous time equations that describe the transfer function between the input voltage $v_{in}(t)$ and the network grid $v_{source}(t)$ are the following

$$\frac{di_1(t)}{dt} = \frac{1}{L_{filter}} \cdot \left( v_{in}(t) - v_2(t) \right) \tag{6}$$

$$\frac{dv_2(t)}{dt} = \frac{1}{\left( C_{filter} + C_{ac\_cable} \right)} \cdot \left( i_1(t) - i_3(t) \right) \tag{7}$$

$$\frac{di_3(t)}{dt} = \frac{1}{L_{fac\_cable}} \cdot \left( v_2(t) - i_3(t) \cdot R_{ac\_cable} - v_3(t) \right) \tag{8}$$

$$\frac{dv_3(t)}{dt} = \frac{i_3(t) - i_4(t)}{C_{ac\_cable}} \tag{9}$$

$$\frac{di_4(t)}{dt} = \frac{v_3(t)}{L_{TR} + L_{source}} \tag{10}$$

The electric parameters related to the capacitive coupling model of Fig. 8 are shown in Table 2. In Fig. 9a, the ground voltage measurement is shown while in Fig. 9b the FFT analysis for this waveform is shown. It is observed that the harmonics components near the switching frequency are considerably higher than the fundamental component. Harmonics components 70 (3500 Hz) is 575% of fundament component magnitude which is 3.05 V. That means that harmonic 70 has a magnitude of 17.54 V, as shown in Fig. 9a. Moreover, the multiples of the switching frequencies are also considerable respect to the fundamental component, as shown in Fig. 9b, where the harmonic component 138 (7000 Hz) and 210 (10500 Hz) are approximately 145% and 98%, respectively, of the fundamental component magnitude.

The ground current waveform measured at the wind farm is shown in Fig. 10a, and the FFT analysis concerning this waveform is performed in Fig. 10b. Consistently with the voltage waveform, the dominant harmonic component in the ground current fits the switching frequency of the converter. That is harmonic component 68 with 503% of the fundamental component magnitude which is 168 mA. Thus, the magnitude of harmonic 68 is 844.9 mA.

| Element | Parameter | Value |
|---------|-----------|-------|
| Wind generator | Operation voltage | 3.5 kV |
|  | Operation frequency | 50 Hz |
|  | Nominal power | 1400 kVA |
|  | Stator winding resistance | 0.01196 pu |
|  | Stator leakage reactance | 0.1966 pu |
| Full converter | Nominal power | 1800 kVA |
|  | Switching frequency | 3500 Hz |
|  | Topology | 6 pulses |
|  | Capacitive coupling | 0.8 uF |
| Filter | Q factor | 10 |
|  | Cut-off frequency | 1000 Hz |
|  | Nominal power | 530 kVA |
| Underground cable | Positive sequence impedance | $0.09015 + j\,0.0426\ \Omega/\text{km}$ |
|  | Zero sequence impedance | $0.0914 + j\,0.03446\ \Omega/\text{km}$ |
|  | Zero sequence susceptance | 0.327 mS/km |
| Power grid | Thevenin voltage | 3.5 kV |
|  | Thevenin inductance | 0. 231 mH |

Table 2. Electric parameters for the wind farm capacitive grounding model.

The multiples of the switching frequencies are also significant, as shown in Fig. 10b, however harmonic component 140 (7000 Hz) appears higher than in the ground voltage waveform near to 200% while harmonic 210 (10500 Hz) is less dominant, 56% but still high enough in comparison with the fundamental component.

These simulation results indicate that ground current in wind farms can be considerable according to the values expressed in (IEEE 80-2000, 2000) for the range of frequencies expressed at Fig. 10a. Therefore, care is then needed to ensure that ground current is within safe limits of work.

This issue is one of the most significant advantages of considering capacitive coupling models for wind farms, which allows implementing further corrective actions to mitigate the adverse effect of ground current over safe conditions of work.

The capacitive coupling model detects the expected resonant frequency of the wind farm at 11.0 kHz with an impedance magnitude Z of 77.8 Ω while simplified models does not detect a resonant frequency for this wind farm configuration, as shown in Fig. 11.

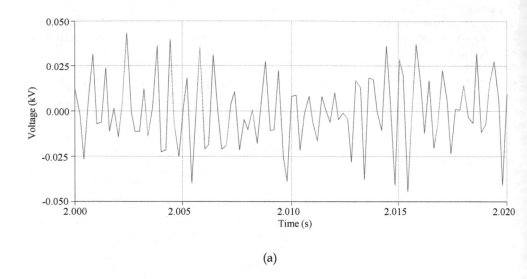

(a)

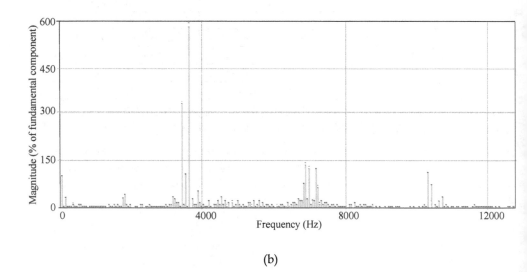

(b)

Fig. 9. Simulation result of the capacitive coupling model: (a) voltage waveform between wind farm electric circuit and grounding system and (b) FFT analysis of the voltage waveform obtained.

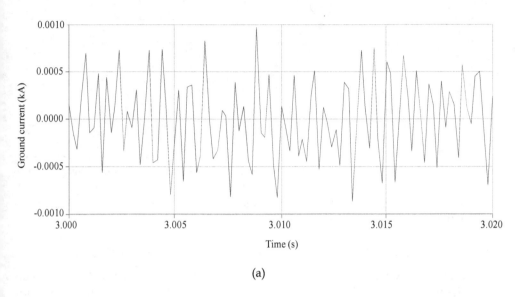

(a)

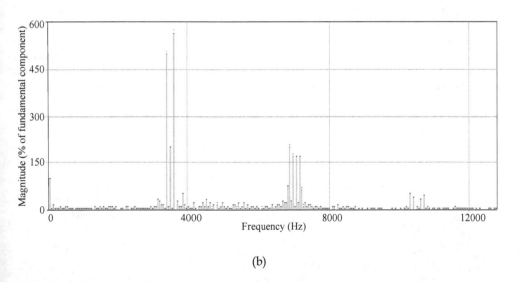

(b)

Fig. 10. Simulation result of the capacitive coupling model: (a) waveform between wind farm electric circuit and ground and (b) FFT analysis of the ground current obtained.

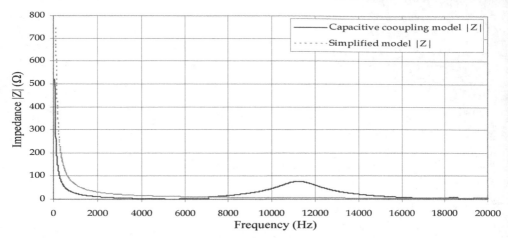

Fig. 11. Resonance frequency of the wind farm model without considering capacitive coupling (dashed line) and with capacitive couplings (solid line).

## 4. Impact on distribution networks of DG ground current contribution

The distribution network considering DG, shown in Fig. 12, has been modelled to analyze the effects of wind farms and PV solar installations ground current contribution to the network. The DG is based on capacitive coupling models of a 1 MW PV solar installation and a 1.4 MW wind farm with the electric parameters shown in Table 1 and Table 2, respectively.

This distribution network feeds two loads through a multi-terminal ring topology. These loads are connected to bus 2 and 5 with a rated power of 500+ $j$ 25 kVA each one.

In steady state conditions, the wind farm generates a total active power of 1370 kW, and the PV solar installation delivered 940 kW to the distribution network. To analyse the capacitive coupling effect over the ground current in DG systems, it has been noticed the voltage and current waveforms seen at node 5 through the capacitive coupling of the line.

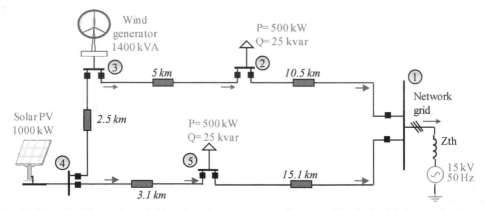

Fig. 12. Distribution network based on capacitive coupling model of wind farms and solar installations.

The electric parameters of the network grid are shown in Table 3.

| Element | Parameter | Value |
|---|---|---|
| Power grid | Thevenin voltage | 15 kV |
| | Thevenin inductance | 17.938 Ω |
| | Shortcircuit power | 12.54 MVA |
| Underground cable | Positive sequence impedance | $0.6969 + j\,0.492\ \Omega/km$ |
| | Zero sequence impedance | $5.945 + j\,7.738\ \Omega/km$ |
| | Zero sequence susceptance | $2.13\ \mu S/km$ |

Table 3. Electric parameters of the network grid.

In node 5, the phase voltage waveform meets the standard regulation of harmonic distortion (THD=5.4%) with a fundamental component of 8.72 kV, as shown in Fig. 13.

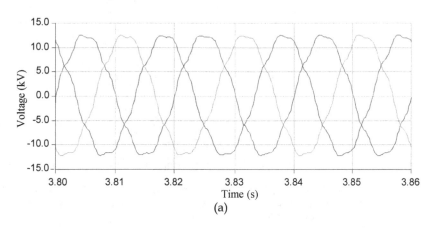

(a)

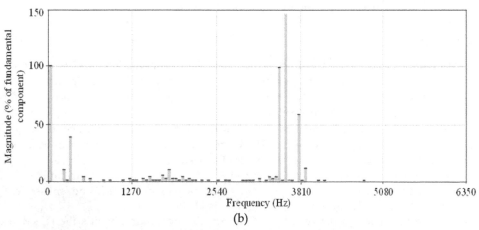

(b)

Fig. 13. Simulation result of the distribution network: (a) phase voltage waveform and (b) FFT analysis of the waveform obtained, at node 5.

Although voltage waveform meets standard regulations, it has been observed an important ground current contribution through the admittance of the underground cables. The ground voltage waveform has a considerable magnitude with peaks reaching 7 V, as shown in Fig. 14. Likewise, the ground current measurement due to the capacitive coupling of these underground cables is also significant as shown in Fig. 15.

The fundamental component of the current waveform is 313 mA, and the THD of this waveform is 190.78%. The most predominant harmonic components are harmonic 72 with 145.22% of the fundamental component, followed by harmonic 70 and 76 with 98.29% and 58.75%, respectively, as shown in Fig. 15a.

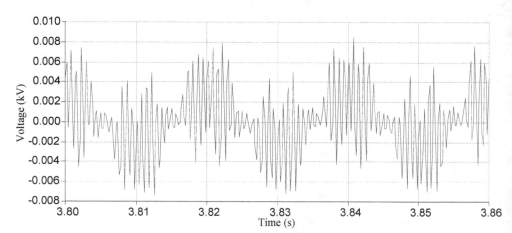

Fig. 14. Simulation result of the distribution network ground voltage waveform at node 5.

These observations point to the importance of controlling the capacitive coupling in load installations connected to networks with DG. Otherwise, end users equipments can be exposed to malfunctioning and lifetime reduction due to the capacitive ground current. Moreover, GPR can reach values of unsafe work conditions.

## 5. Conclusions

The capacitive coupling models lead to an accurate approximation to the response of distribution network against the frequency spectrum imposed by the switching action of the converters at DG. This approximation is not feasible using simplified models because of the bandwidth limitation for high frequencies.

According to the distribution network under study, a high ground current contribution to grid provided by DG has been detected. Therefore, some preventive actions can be applied to network design stage in order to solve this problem, such as:

-   Connection of the PV array to the grounding systems by means of an inductor. The latter element represents high impedance for harmonics current and subsequently reduces the capacitive ground current in the installation.
-   Insertion of capacitors between the DC terminals and ground avoids the injection of harmonic current to the PV array, as shown in Fig. 13b, and thereby the noise level and GPR between PV modules and ground is minimized.

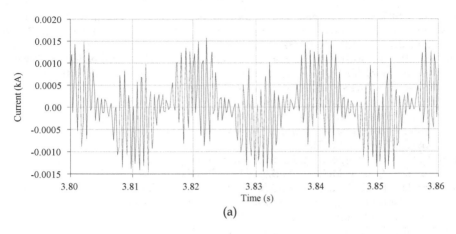

(a)

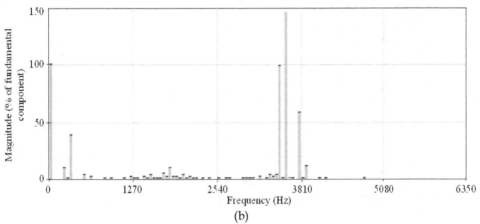

(b)

Fig. 15. Simulation result of the distribution network: (a) current waveform and (b) FFT analysis of the waveform obtained, at node 5.

- Adjustment of the firing pulses frequencies and control strategies to reduce or avoid resonance and capacitive currents by analyzing the ground current with the proposed model.
- Insertion of high-pass filters on the distribution network to avoid end users equipments to be exposed to a high amount of ground current.

## 6. References

Bellini A., Bifaretti S., Iacovone V. & Cornaro C. (2009). Simplified model of a photovoltaic module, *International Conference on Applied Electronics*, pp. 47–51, ISBN 978-80-7043-781-0, Pilsen, Bohemia, Czech Republic, Sept. 9-10, 2009

Chayawatto N., Kirtikara K., Monyakul V., Jivacate, C. & Chenvidhya D. (2009). DC/AC switching converter modeling of a PV grid-connected system under islanding phenomena, *Renewable Energy*, Vol. 34, No. 12, (2009), pp. 2536–44, ISSN 0960-1481

Chicco G., Schlabbach J. & Spertino F., (2009). Experimental assessment of the waveform distortion in grid-connected photovoltaic installations, *Solar Energy*, Vol. 83, No. 1, pp.1026–39, (2009), ISSN 0038-092X

Comech, M.P., García-Gracia, M., Borroy, S., Villén, M.T. (2010). Protection in Distributed Generation, In: *Distributed Generation*, D. N. Gaonkar, pp. 289-310, In-Teh, ISBN 978-953-307-046-9, Olajnica, Vukovar, Croatia

Conroy, J. & Watson, R. (2009). Aggregate modelling of wind farms containing full-converter wind turbine generators with permanent magnet synchronous machines: transient stability studies, *IET Renewable Power Generation*, Vol. 3, No. 1, (2009), pp. 39–52, ISSN 1752-1424

Dugan, R. C., McGranaghan, M. F., Santoso, S. & Wayne Beaty, H. (2002). *Electrical Power Systems Quality*, McGraw-Hill, (2nd Ed.), ISBN 0-07-138622-X, New York, USA

García-Gracia, M., El Halabi, N., Khodr, H.M. & Sanz, J. F. (2010). Improvement of large scale solar installation model for ground current analysis, *Applied Energy*, Vol. 87, No. 11, (2010), pp. 3467-3474, ISSN 0306-2619

IEC 60479-2 (1987). Effect of current passing through human body. Part II: special aspects. *The International Electrotechnical Commission*, 1987.

IEC Std. 61000-4-7 (2002). Electromagnetic compatibility (EMC). Part 4-7: Testing and measurement techniques. General guide on harmonics and interharmonics measurements and instrumentation for power supply and equipment connected thereto, *International Electrotechnical Commission*, (2002)

IEC Std. 61400-21 (2008), Wind turbines. Part 21: Measurement and assessment of power quality characteristics of grid connected wind turbine, *International Electrotechnical Commission*, (2008).

IEEE Std 519-1992 (1992). (1992). *IEEE recommended practices and requirements for harmonic control in electrical power systems*, Power Engineering Society, 1992

IEEE Std. 80-2000 (2000). *IEEE guide for safety in AC substation grounding*, IEEE Power Engineering Society, 2000

Iliceto A. & Vigotti R. (1998). The largest PV installation in Europe: perspectives of multimegawatt PV, *Renewable Energy*, Vol. 15, No. 1-4, (1998), pp. 48–53, ISSN 0960-1481

Kim S-K., Jeon J-H., Cho C-H., Kim E-S., Ahn J-B. (2009). Modelling and simulation of a grid-connected PV generation system for electromagnetic transient analysis, *Solar Energy*, Vol. 83, No. 5, (2009), pp. 664–78, ISSN 0038-092X

Luna, A., De Araujo, F., Santos, D., Rodriguez, P., Watanabe, E. & Arnaltes, S. (2011). Simplified Modeling of a DFIG for Transient Studies in Wind Power Applications, *IEEE Transactions on Industrial Electronics*, Vol. 58, No. 1, (2011), pp. 9–20, ISSN 0278-0046

Sukamongkol SCY & Ongsakul W. (2002), A simulation model for predicting the performance of a solar photovoltaic system with alternating current loads, *Renewable Energy*, Vol. 27, No. 2, pp. 237–58, (2002), ISSN 0960-1481

Villalva M., Gazoli J, Filho E. (2009), Comprehensive approach to modeling and simulation of photovoltaic arrays, *IEEE Transactions Power Electronics*, Vol. 24, No. 5, pp. 1198–208, (2009), ISSN 0885-8993

Zhou, W., Lou, C., Li, Z., Lu, L., & Yang, H. (2010). Current status of research on optimum sizing of stand-alone hybrid solar-wind power generation systems, *Applied Energy*, Vol. 87, No. 2, pp. 380–389, ISSN 0306-2619

# 10

# Power Quality Problems Generated by Line Frequency Coreless Induction Furnaces

Angela Iagăr
*Politechnica University Timişoara*
*Romania*

## 1. Introduction

The increased problems in power networks impose to identify the sources of power quality deterioration. The most important parameters which affect power quality are harmonics, voltage instability and reactive power burden (Arrillaga et al., 2000). They cause low system efficiency, poor power factor, cause disturbance to other consumers and interference in the nearly communication networks (Lattarulo, 2007; De la Rosa, 2006; Muzi, 2008).

In induction melting is noticed mainly the efficiency, high heating rate and the reduced oxidation level of the processed material, the improved work conditions and the possibility of an accurate control of the technological processes (Rudnev et al., 2002).

Induction heating equipments do not introduce dust and noise emissions in operation, but cause power quality problems in the electric power system (Nuns et al., 1993).

Induction-melt furnaces supplies by medium frequency converters generate fixed and variable frequency harmonics. Both current and voltage-fed inverters generate harmonics back into power lines in the process of rectifying AC to DC (EPRI, 1999).

Harmonics flowing in the network causing additional losses and decreasing the equipments lifetime. Also, the harmonics can interfere with control, communication or protection equipments (Arrillaga et al., 2000; George & Agarwal, 2008).

In addition to the harmonics that are normally expected from different pulse rectifiers, large furnaces operating at a few hundred hertz can generate interharmonics (EPRI, 1999). Interharmonics can overload power system capacitors, introduce noise into transformers, cause lights to flicker, instigate UPS alarms, and trip adjustable-speed drives.

High-frequency systems, which operate at greater than 3 kHz are relatively small and limited to special applications. Electromagnetic pollution produced by the operation of these equipments is small.

The induction furnaces supplied at line frequency (50 Hz) are of high capacity and represent great power consumers.

Being single-phase loads, these furnaces introduce unbalances that lead to the increasing of power and active energy losses in the network. In case of channel furnaces it was found the presence of harmonics in the current absorbed from the power supply network. These harmonics can be determined by the non-sinusoidal supply voltages or the load's nonlinearity, owed to the saturation of the magnetic circuit (Nuns et al., 1993).

This chapter presents a study about power quality problems introduced by the operation of line frequency coreless induction furnaces. The specialty literature does not offer detailed information regarding the harmonic distortion in the case of these furnaces.

## 2. Electrical installation of the induction-melt furnace

The analyzed coreless induction furnace has 12.5 t capacity of cast-iron; the furnace is supplied from the three-phase medium-voltage network (6 kV) through a transformer in $\Delta/Y$ connection, with step-variable voltage. Load balancing of the three-phase network is currently achieved by a Steinmetz circuit, and the power factor correction is achieved by means of some step-switching capacitor banks (fig.1).

In the electric scheme from fig. 1: $Q_1$ is an indoor three-poles disconnector, type STIm–10–1250 (10 kV, 1250 A), $Q_2$ is an automatic circuit-breaker OROMAX (6 kV, 2500 A), T is the furnace transformer (2625 kVA; 6/1.2 kV), $K_1$ is a contactor (1600 A), (1) is the Steinmetz circuit used to balance the line currents, (2) is the power factor compensation installation, $TC_{1m}$, $TC_{2m}$, $TC_{3m}$ (300/5 A) and $TC_1$, $TC_2$, $TC_3$ (1600/5 A) are current transformers, $TT_{1m}$ (6000/100 V), $TT_1$ (1320/110 V) are voltage transformers, and M is the flexible connection of the induction furnace CI.

Within the study the following physical aspects were taken into account:

- induction heating of ferromagnetic materials involve complex and strongly coupled phenomena (generating of eddy currents, heat transfer, phase transitions and mechanical stress of the processed material);
- the resistivity of cast-iron increases with temperature;
- the relative magnetic permeability of the cast-iron changes very fast against temperature near to the Curie point (above the Curie temperature the cast-iron becomes paramagnetic).

As consequence, will be present the influence of the following factors upon the energetic parameters of the installation: furnace charge, furnace supply voltage, load balancing installation and the one of power factor compensation.

## 3. Measured signals in electrical installation of the induction furnace

The measurements have been made both in the secondary (Low Voltage Line - LV Line) and in the primary (Medium Voltage Line - MV Line) of the furnace transformer, using the CA8334 three-phase power quality analyzer. CA8334 gave an instantaneous image of the main characteristics of power quality for the analyzed induction furnace. The main parameters measured by the CA8334 analyser were: TRMS AC phase voltages and TRMS AC line currents; peak voltage and current; active, reactive and apparent power per phase; harmonics for voltages and currents up to the 50th order (CA8334, technical handbook, 2007).

CA8334 analyser provide numerous calculated values and processing functions in compliance with EMC standards in use (EN 50160, IEC 61000-4-15, IEC 61000-4-30, IEC 61000-4-7, IEC 61000-3-4).

The most significant moments during the induction melting process of the cast-iron charge were considered:

- cold state of the charge (after 15 minutes from the beginning of the heating process);
- intermediate state (after 5 hours and 40 minutes from the beginning of the heating process);
- the end of the melting (after 8 hours from the beginning of the heating process).

Further are presented the waveforms and harmonic spectra of the phase voltages and line currents measured during the heating of the charge (Iagăr et al, 2009).

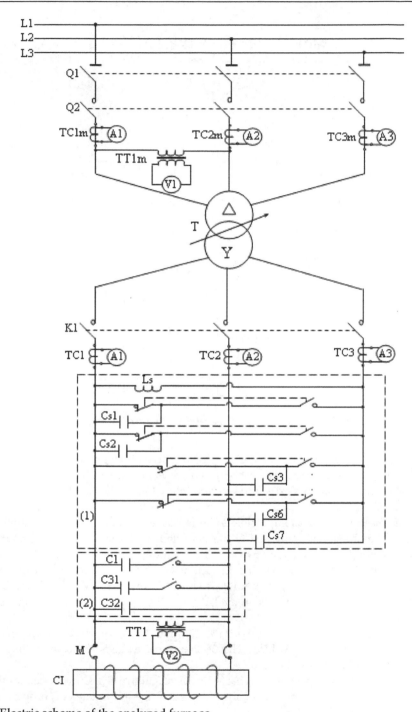

Fig. 1. Electric scheme of the analyzed furnace

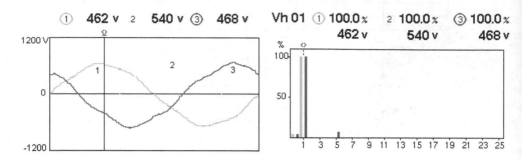

Fig. 2. Waveforms and harmonic spectra of the phase voltages in the cold state of the charge (LV Line)

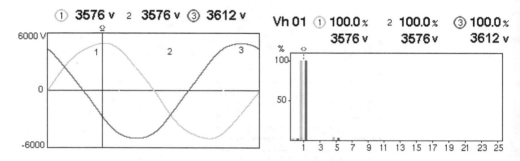

Fig. 3. Waveforms and harmonic spectra of the phase voltages in the cold state of the charge (MV Line)

In the first heating stage, the electromagnetic disturbances of the phase voltages on LV Line and on MV Line are very small. The 5th harmonic does not exceed the compatibility limit, but the voltage interharmonics exceed the compatibility limits (IEC 61000-3-4, 1998; IEC/TR 61000-3-6, 2005).

On MV Line the current $I_2$ was impossible to be measured because the CA8334 three-phase power quality analyser was connected to the watt-hour meter input. The watt-hour meter had three voltages ($U_{12}$, $U_{23}$, $U_{31}$) and two currents ($I_1$ and $I_3$).

Waveform distortion of the currents in cold state is large (fig. 4, 5). At the beginning of the cast-iron heating the 3rd, 5th, 7th, 9th, 11th, 13th, 15th harmonics and even harmonics (2nd, 4th, 6th, 8th) are present in the currents on the LV Line. The 5th and 15th harmonics exceed the compatibility limits (IEC 61000-3-4, 1998).

In the cold state the 2nd, 3rd, 5th, 7th, 9th, 11th, 13th and 15th harmonics are present in the currents absorbed from the MV Line. The 5th harmonic exceeds the compatibility limits (IEC/TR 61000-3-6, 2005).

In the intermediate state, part of the charge is heated above the Curie temperature and becomes paramagnetic, and the rest of the charge still has ferromagnetic properties. The furnace charge is partially melted.

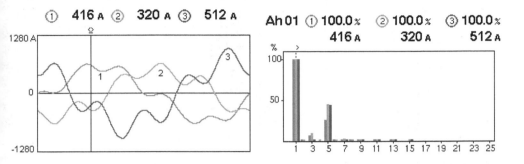

Fig. 4. Waveforms and harmonic spectra of the line currents in the cold state of the charge (LV Line)

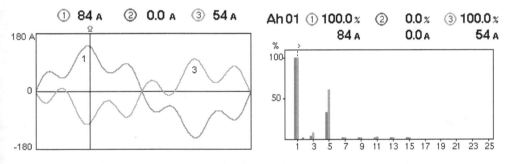

Fig. 5. Waveforms and harmonic spectra of the line currents in the cold state of the charge (MV Line)

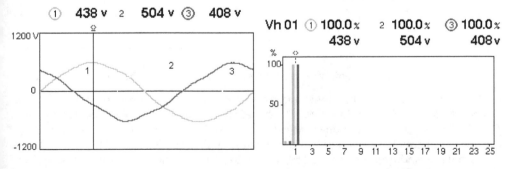

Fig. 6. Waveforms and harmonic spectra of the phase voltages in the intermediate state (LV Line)

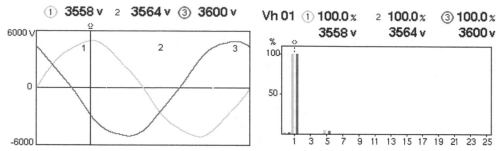

Fig. 7. Waveforms and harmonic spectra of the voltages in the intermediate state (MV Line)

In the intermediate state of the charge, the voltage interharmonics exceed the compatibility limits. The 5th harmonic do not exceeds the compatibility limits.

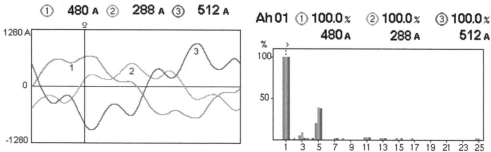

Fig. 8. Waveforms and harmonic spectra of the currents in the intermediate state (LV Line)

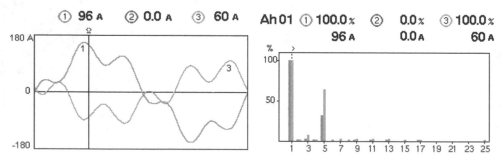

Fig. 9. Waveforms and harmonic spectra of the currents in the intermediate state (MV Line)

In the intermediate state, harmonic spectra of the currents absorbed from the LV Line present the 3rd, 5th, 7th, 11th, 13th, 15th, 17th, 25th harmonics and even harmonics (2nd, 4th, 8th). The 5th, 15th, 17th and 25th harmonics exceed the compatibility limits (IEC 61000-3-4, 1998).

On MV Line, harmonic spectra of the currents present the 3rd, 5th, 7th, 9th, 11th, 13th, 15th, 17th, 25th harmonics and even harmonics (2nd, 4th, 6th, 8th). The 5th and 25th harmonics exceed the compatibility limits (IEC/TR 61000-3-6, 2005).

After 8 hours from the beginning of the heating process the furnace charge is totally melted, being paramagnetic.

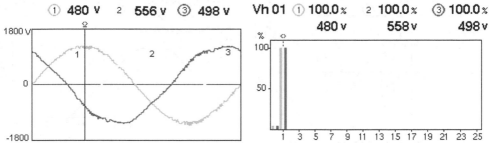

Fig. 10. Waveforms and harmonic spectra of the phase voltages at the end of the melting process (LV Line)

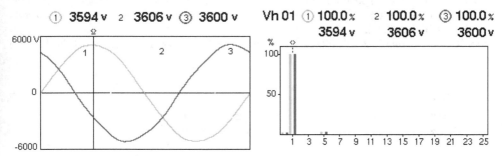

Fig. 11. Waveforms and harmonic spectra of the phase voltages at the end of the melting process (MV Line)

At the end of the melting process, the electromagnetic disturbances of the phase voltages are very small. Voltage interharmonics exceed the compatibility limits. The 5th harmonic is within compatibility limits (IEC 61000-3-4, 1998; IEC/TR 61000-3-6, 2005).

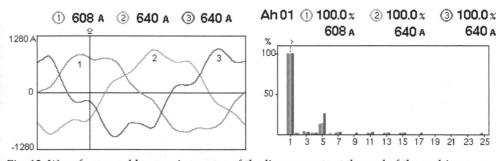

Fig. 12. Waveforms and harmonic spectra of the line currents at the end of the melting process (LV Line)

Waveform distortion of the currents at the end of the melting process is smaller than in cold state, or intermediate state. On LV Line, harmonic spectra of the currents show the presence of 3rd, 5th, 7th, 9th, 11th, 13th, 15th, 17th, 25th harmonics and even harmonics (2nd, 4th, 6th). The 5th, 15th and 25th harmonics exceed the compatibility limits (IEC 61000-3-4, 1998).

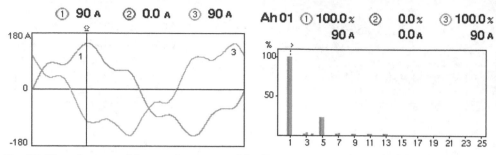

Fig. 13. Waveforms and harmonic spectra of the line currents at the end of the melting process (MV Line)

On MV Line, harmonic spectra of the currents show the presence of 3rd, 4th, 5th, 7th, 9th, 11th, 13th harmonics at the end of the melting. The 5th harmonic exceeds the compatibility limits (IEC/TR 61000-3-6, 2005).

Fig.14-16 show the values of voltage and current unbalance on LV Line, in all the heating stages.

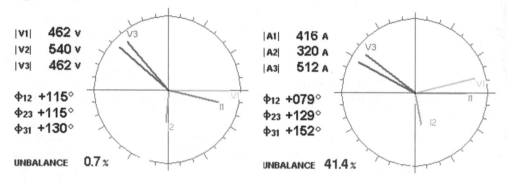

Fig. 14. Unbalance of the phase voltages and line currents in the cold state of the charge (LV Line)

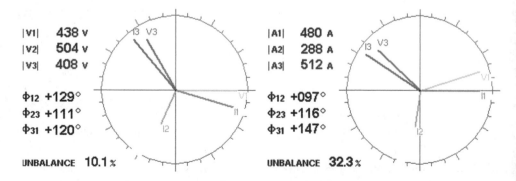

Fig. 15. Unbalance of the phase voltages and line currents in the intermediate state of the charge (LV Line)

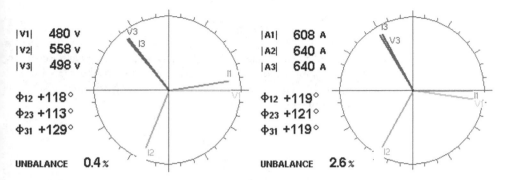

| $|V_1|$ | 480 v |
| $|V_2|$ | 558 v |
| $|V_3|$ | 498 v |

$\Phi_{12}$ +118°
$\Phi_{23}$ +113°
$\Phi_{31}$ +129°

**UNBALANCE** 0.4%

| $|A_1|$ | 608 A |
| $|A_2|$ | 640 A |
| $|A_3|$ | 640 A |

$\Phi_{12}$ +119°
$\Phi_{23}$ +121°
$\Phi_{31}$ +119°

**UNBALANCE** 2.6%

Fig. 16. Unbalance of the phase voltages and line currents at the end of the melting process (LV Line)

Voltage unbalance exceeds the permitted values in intermediate state. Current unbalance is very large in the cold state and decreases as the furnace charge is melting down.

## 4. The values computed by the CA8334 analyser

The values computed by the CA8334 analyser are: total harmonic distortion of voltages and currents, distortion factor of voltages and currents, K factor for current, voltage and current unbalance, power factor and displacement factor, extreme and average values for voltage and current, peak factors for current and voltage (CA8334, technical handbook, 2007). Mathematical formulae used to compute the total harmonic distortion (THD) of voltages and currents are:

$$VTHD_i = \frac{\sqrt{\sum_{n=2}^{50}(V_{harm\ ni})^2}}{V_{harm\ 1i}} \cdot 100 \tag{1}$$

$$ITHD_i = \frac{\sqrt{\sum_{n=2}^{50}(I_{harm\ ni})^2}}{I_{harm\ 1i}} \cdot 100 \tag{2}$$

V represents the phase voltage, I represents the line current, i represents the phase (i = 1, 2, 3) and n represents the order of harmonics.
Distortion factor (DF) of voltages and currents are computed by the formulae:

$$VDF_i = \frac{\sqrt{\frac{1}{2}\sum_{n=2}^{50}(V_{harm\ ni})^2}}{VRMS_i} \cdot 100 \tag{3}$$

$$IDF_i = \frac{\sqrt{\frac{1}{2}\sum_{n=2}^{50}(I_{harm\ ni})^2}}{IRMS_i} \cdot 100 \tag{4}$$

VRMS and IRMS represent the root mean square values (RMS values or effective values) for phase voltage and line current, computed over 1 second, and i represents the phase (i = 1, 2, 3). **K factor (KF)** is a weighting of the harmonic load currents according to their effects on transformer heating. **K factor** for current is computed by relation:

$$IKF_i = \frac{\sum_{n=1}^{50}n^2 \cdot (I_{harm\ ni})^2}{\sum_{n=1}^{50}(I_{harm\ ni})^2} \tag{5}$$

In the above relation I represents the line current, i represents the phase (i = 1, 2, 3) and n represents the order of harmonics. A K factor of 1 indicates a linear load (no harmonics); a higher K factor indicates the greater harmonic heating effects.

The unbalanced three-phase systems of voltages (or currents) can be reduce into three balanced systems: the positive (+), negative (-) and zero (0) sequence components.

The positive voltage True RMS and the negative voltage True RMS are given by the relations:

$$VRMS_+ = \frac{V_1 + aV_2 + a^2V_3}{3} \tag{6}$$

$$VRMS_- = \frac{V_1 + a^2V_2 + aV_3}{3} \tag{7}$$

where $V_1, V_2, V_3$ represent the phase voltages (using simplified complex) and $a = e^{j\frac{2\pi}{3}}$ is the complex operator.

The positive current True RMS and the negative current True RMS are given by the relations:

$$IRMS_+ = \frac{I_1 + aI_2 + a^2I_3}{3} \tag{8}$$

$$IRMS_- = \frac{I_1 + a^2I_2 + aI_3}{3} \tag{9}$$

where $I_1, I_2, I_3$ represent the line currents (using simplified complex).

Voltage and current unbalances (unb) are:

$$V_{unb} = \frac{|VRMS_-|}{|VRMS_+|} \cdot 100 \tag{10}$$

$$I_{unb} = \frac{|IRMS_-|}{|IRMS_+|} \cdot 100 \qquad (11)$$

**Power factor (PF) and displacement factor (DPF)** are computed by relations:

$$PF_i = \frac{P_i}{S_i} \qquad (12)$$

$$DPF_i = \cos\varphi_i \qquad (13)$$

$P_i$ [W] and $S_i$ [VA] represent the active power and the apparent power per phase (i = 1, 2, 3); $\varphi_i$ is the phase difference between the fundamental current and voltage, and i represents the phase.

Mathematical formulae used to compute the **peak factors (CF) for current and phase voltage** are:

$$VCF_i = \frac{\max(Vpp_i, |Vpm_i|)}{\sqrt{\frac{1}{N} \cdot \sum_{n=0}^{N-1} (V(n)_i)^2}} \qquad (14)$$

$$ICF_i = \frac{\max(Ipp_i, |Ipm_i|)}{\sqrt{\frac{1}{N} \cdot \sum_{k=0}^{N-1} (I(k)_i)^2}} \qquad (15)$$

In the relations (14), (15): Vpp is the PEAK+ of the phase voltage; Vpm is the PEAK- of the phase voltage; Ipp is the PEAK+ of the line current; Ipm is the PEAK- of the line current; i represents the phase (i = 1, 2, 3); N represents the number of the samples per period (between two consecutive zeros).

Peak values (PEAK+/PEAK-) for voltage (or current) represent the maximum/minimum values of the voltage (or current) for all the samples between two consecutive zeros. For a sinusoidal signal, the peak factor is equal to $\sqrt{2}$ (1.41). For a non-sinusoidal signal, the peak factor can be either greater than or less than $\sqrt{2}$. In the latter case, the peak factor signals divergent peak values with respect to the RMS value.

MIN/MAX values for voltage (or current) represent the minimum/maximum values of the half-period RMS voltage (or current). Average values (AVG) for voltage and current are computed over 1 second.

Tables 1-25 show the values computed by the CA8334 analyser on LV Line and on MV Line.

| Heating moment | VTHD$_1$[%] | VTHD$_2$[%] | VTHD$_3$[%] |
|---|---|---|---|
| Cold state | 0 | 4 | 5.4 |
| Intermediate state | 0 | 3.8 | 3.8 |
| End of melting process | 0 | 0 | 6.3 |

Table 1. Total harmonic distortion THD [%] for phase voltages (LV Line)

| Heating moment | VTHD$_1$[%] | VTHD$_2$[%] | VTHD$_3$[%] |
|---|---|---|---|
| Cold state | 2.2 | 0 | 1.7 |
| Intermediate state | 3.4 | 0 | 3.1 |
| End of melting process | 1.9 | 0 | 1.7 |

Table 2. Total harmonic distortion THD [%] for phase voltages (MV Line)

THD of the phase voltages do not exceed the compatibility limits in all the heating stages. The values of VTHD on MV Line are higher than the values of VTHD on LV Line.

| Heating moment | ITHD$_1$[%] | ITHD$_2$[%] | ITHD$_3$[%] |
|---|---|---|---|
| Cold state | 26.5 | 43 | 42 |
| Intermediate state | 20.1 | 39 | 35.5 |
| End of melting process | 14.9 | 16.7 | 30.3 |

Table 3. Total harmonic distortion THD [%] for line currents (LV Line)

| Heating moment | ITHD$_1$[%] | ITHD$_3$[%] |
|---|---|---|
| Cold state | 31 | 57.5 |
| Intermediate state | 34.3 | 68.7 |
| End of melting process | 22.7 | 24.3 |

Table 4. Total harmonic distortion THD [%] for line currents (MV Line)

ITHD exceed the limits permitted by norms in all the analyzed situations. The values of ITHD are higher on MV Line versus LV Line. Because THD of line currents exceed 20%, this indicates a significant electromagnetic pollution produced by the furnace in MV network.

| Heating moment | VDF$_1$[%] | VDF$_2$[%] | VDF$_3$[%] |
|---|---|---|---|
| Cold state | 0 | 0 | 0 |
| Intermediate state | 0 | 0 | 0 |
| End of melting process | 0 | 0 | 5.5 |

Table 5. Distortion factor DF [%] of phase voltages (LV Line)

| Heating moment | VDF$_1$[%] | VDF$_2$[%] | VDF$_3$[%] |
|---|---|---|---|
| Cold state | 1.9 | 0 | 1.4 |
| Intermediate state | 3.4 | 0 | 3 |
| End of melting process | 1.9 | 0 | 1.4 |

Table 6. Distortion factor DF [%] of phase voltages (MV Line)

Distortion factor of phase voltages is very small during the heating process of cast-iron charge. In all situations, distortion factor of phase voltages is smaller than total harmonic distortion.

| Heating moment | IDF$_1$[%] | IDF$_2$[%] | IDF$_3$[%] |
|---|---|---|---|
| Cold state | 21.7 | 46.2 | 32 |
| Intermediate state | 19.3 | 38.1 | 33.4 |
| End of melting process | 15.6 | 14.8 | 27.9 |

Table 7. Distortion factor DF [%] of line currents (LV Line)

| Heating moment | IDF$_1$[%] | IDF$_3$[%] |
|---|---|---|
| Cold state | 23.9 | 43.2 |
| Intermediate state | 32.2 | 61.8 |
| End of melting process | 22.7 | 23.4 |

Table 8. Distortion factor DF [%] of line currents (MV Line)

The values of distortion factor of line currents are very high during the heating process (Table 7 and Table 8). The values of IDF are higher on MV Line versus LV Line.

| Heating moment | IKF$_1$ | IKF$_2$ | IKF$_3$ |
|---|---|---|---|
| Cold state | 2.02 | 6.07 | 3.52 |
| Intermediate state | 1.88 | 4.8 | 4.02 |
| End of melting process | 1.59 | 1.58 | 2.93 |

Table 9. K factor KF [-] of line currents (LV Line)

| Heating moment | IKF$_1$ | IKF$_3$ |
|---|---|---|
| Cold state | 2.51 | 5.59 |
| Intermediate state | 3.54 | 8.7 |
| End of melting process | 2.21 | 2.27 |

Table 10. K factor KF [-] of line currents (MV Line)

K factor is greater than unity in all the heating stages. The values of K factor in the cold state and in the intermediate state are very high. This indicates the significant harmonic current content. K factor decrease at the end of the melting. Harmonics generate additional heat in the furnace transformer. If the transformer is non-K-rated, overheat possibly causing a fire, also reducing the life of the transformer.

| Heating moment | PF | | | DPF | | |
|---|---|---|---|---|---|---|
| | 1 | 2 | 3 | 1 | 2 | 3 |
| Cold state | 0.96 | 0.84 | 0.93 | 0.98 | 0.93 | 0.99 |
| Intermediate state | 0.93 | 0.88 | 0.92 | 0.95 | 0.97 | 0.98 |
| End of melting process | 0.97 | 0.97 | 0.96 | 0.99 | 0.99 | 0.99 |

Table 11. PF [-] and DPF [-] per phase (1, 2, 3) on LV line

PF is less than unity in all the analyzed situations on LV Line. In the cold state and in the intermediate state, PF is less than neutral value (0.92) per phase 2.

| Values | $u_1$ | $u_2$ | $u_3$ |
|--------|-------|-------|-------|
| MAX [V] | 552 | 624 | 558 |
| AVG [V] | 456 | 540 | 468 |
| MIN [V] | 0 | 0 | 0 |
| PEAK+ [V] | 660 | 786 | 678 |
| PEAK- [V] | -672 | -786 | -726 |

Table 12. Extreme and average values for phase voltages in the cold state (LV line)

| Values | $u_1$ | $u_2$ | $u_3$ |
|--------|-------|-------|-------|
| MAX [V] | 4176 | 4182 | 4158 |
| AVG [V] | 3558 | 3564 | 3600 |
| MIN [V] | 0 | 2862 | 2796 |
| PEAK+ [V] | 5034 | 5058 | 5028 |
| PEAK- [V] | -5076 | -5076 | -5046 |

Table 13. Extreme and average values for phase voltages in the cold state (MV Line)

| Values | $u_1$ | $u_2$ | $u_3$ |
|--------|-------|-------|-------|
| MAX [V] | 498 | 570 | 516 |
| AVG [V] | 486 | 564 | 504 |
| MIN [V] | 456 | 540 | 474 |
| PEAK+ [V] | 708 | 828 | 732 |
| PEAK- [V] | -732 | -810 | -768 |

Table 14. Extreme and average values for phase voltages at the end of melting (LV line)

| Values | $u_1$ | $u_2$ | $u_3$ |
|--------|-------|-------|-------|
| MAX [V] | 4140 | 4068 | 4146 |
| AVG [V] | 3594 | 3606 | 3600 |
| MIN [V] | 3558 | 3522 | 3480 |
| PEAK+ [V] | 5052 | 5118 | 5028 |
| PEAK- [V] | -5094 | -5136 | -5046 |

Table 15. Extreme and average values for phase voltages at the end of melting (MV Line)

Tables 12-15 indicate a small unbalance of phase voltages in all the analyzed situations, on LV Line and on MV Line.

| Values | $i_1$ | $i_2$ | $i_3$ |
|--------|-------|-------|-------|
| MAX [A] | 1150 | 732 | 1665 |
| AVG [A] | 416 | 224 | 544 |
| MIN [A] | 0 | 0 | 0 |
| PEAK+ [A] | 608 | 384 | 928 |
| PEAK- [A] | -608 | -384 | -928 |

Table 16. Extreme and average values for line currents in the cold state (LV line)

| Values | $i_1$ | $i_3$ |
|--------|-------|-------|
| MAX [A] | 96 | 60 |
| AVG [A] | 84 | 48 |
| MIN [A] | 0 | 0 |
| PEAK+ [A] | 138 | 90 |
| PEAK- [A] | -138 | -90 |

Table 17. Extreme and average values for line currents in the cold state (MV line)

| Values | $i_1$ | $i_2$ | $i_3$ |
|--------|-------|-------|-------|
| MAX [A] | 1267 | 976 | 1713 |
| AVG [A] | 480 | 288 | 544 |
| MIN [A] | 0 | 0 | 0 |
| PEAK+ [A] | 704 | 512 | 992 |
| PEAK- [A] | -704 | -512 | -992 |

Table 18. Extreme and average values for line currents in the intermediate state (LV line)

| Values | $i_1$ | $i_3$ |
|--------|-------|-------|
| MAX [A] | 324 | 240 |
| AVG [A] | 96 | 60 |
| MIN [A] | 0 | 0 |
| PEAK+ [A] | 162 | 108 |
| PEAK- [A] | -162 | -102 |

Table 19. Extreme and average values for line currents in the intermediate state (MV line)

| Values | $i_1$ | $i_2$ | $i_3$ |
|--------|-------|-------|-------|
| MAX [A] | 672 | 672 | 672 |
| AVG [A] | 608 | 640 | 672 |
| MIN [A] | 544 | 544 | 608 |
| PEAK+ [A] | 896 | 992 | 1088 |
| PEAK- [A] | -896 | -992 | -1056 |

Table 20. Extreme and average values for line currents at the end of melting (LV line)

| Values | $i_1$ | $i_3$ |
|---|---|---|
| MAX [A] | 102 | 102 |
| AVG [A] | 90 | 90 |
| MIN [A] | 90 | 84 |
| PEAK+ [A] | 150 | 150 |
| PEAK- [A] | -150 | -150 |

Table 21. Extreme and average values for line currents at the end of melting (MV line)

The extreme and average values of line currents indicate a large unbalance in the cold state and in intermediate state. At the end of the melting the unbalance of currents is small.

| Heating moment | $VCF_1$ | $VCF_2$ | $VCF_3$ |
|---|---|---|---|
| Cold state | 1.47 | 1.46 | 1.53 |
| Intermediate state | 1.48 | 1.44 | 1.56 |
| End of melting process | 1.45 | 1.47 | 1.49 |

Table 22. Peak factors CF [-] of phase voltages (LV Line)

| Heating moment | $VCF_1$ | $VCF_2$ | $VCF_3$ |
|---|---|---|---|
| Cold state | 1.42 | 1.42 | 1.39 |
| Intermediate state | 1.44 | 1.42 | 1.39 |
| End of melting process | 1.45 | 1.47 | 1.49 |

Table 23. Peak factors CF [-] of phase voltages (MV Line)

Peak factors of phase voltages do not exceed very much the peak factor for sinusoidal signals (1.41) in all the heating stages. This indicates a small distortion of phase voltages.

| Heating moment | $ICF_1$ | $ICF_2$ | $ICF_3$ |
|---|---|---|---|
| Cold state | 1.59 | 1.83 | 1.81 |
| Intermediate state | 1.51 | 1.88 | 1.83 |
| End of melting process | 1.48 | 1.64 | 1.66 |

Table 24. Peak factors CF [-] of line currents (LV Line)

| Heating moment | $ICF_1$ | $ICF_3$ |
|---|---|---|
| Cold state | 1.68 | 1.82 |
| Intermediate state | 1.72 | 1.79 |
| End of melting process | 1.68 | 1.68 |

Table 25. Peak factors CF [-] of line currents (MV Line)

Peak factors of line currents are between 1.48 and 1.88. This indicates that the analyzed furnace is a non-linear load. A high peak factor characterizes high transient overcurrents which, when detected by protection devices, can cause nuisance tripping.

## 5. Recorded parameters in the electrical installation of the induction furnace

The recorded parameters in the electrical installation of analyzed furnace are: RMS values of phase voltages and currents, total harmonic distortion of phase voltages and currents, power factor and displacement factor per phase 1, active power, reactive power and apparent power per phase 1.

Fig.17-21 show the recorded parameters on MV Line, in the first stage of the heating. In the recording period (11:20-12:18), the furnace charge was ferromagnetic.

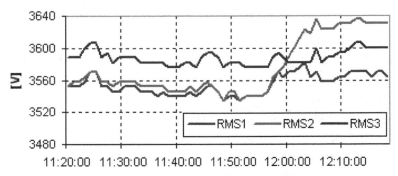

Fig. 17. RMS values of the phase voltages in the cold state (MV Line)

RMS values of phase voltages in the cold state indicate a small unbalance of the load. THD of phase voltages are within compatibility limits in the first stage of the heating process.

The RMS values of line currents show a poor balance between the phases. The Steinmetz circuit is not efficient for load balancing in this stage of the melting process.

THD of line currents have values of 20%...70%, and exceed very much the compatibility limits during the recording period. This indicates a significant harmonic pollution with a risk of temperature rise.

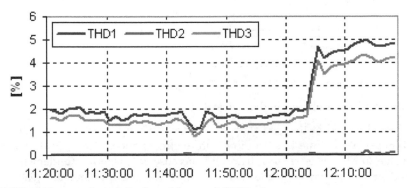

Fig. 18. THD of phase voltages in the cold state (MV Line)

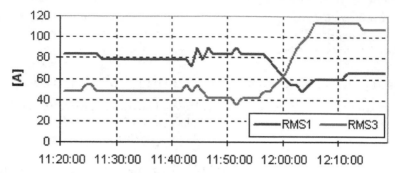

Fig. 19. RMS values of the currents in the cold state (MV Line)

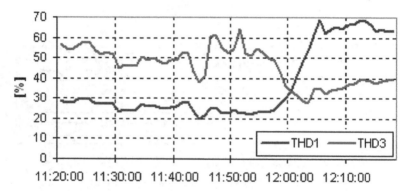

Fig. 20. THD of line currents in the cold state (MV Line)

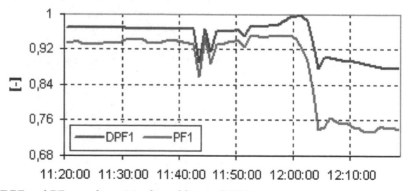

Fig. 21. DPF and PF per phase 1 in the cold state (MV Line)

In the recorded period of the cold state, power factor (PF) per phase 1 and displacement factor (DPF) per phase 1 are less than unity; in the time period 12:00 - 12:18 PF is less than neutral value (0.92). PF is smaller than DPF because PF includes fundamental reactive power and harmonic power, while DPF only includes the fundamental reactive power caused by a phase shift between voltage and fundamental current.

Fig.22-29 show the recorded parameters in the intermediate state of the heating. The furnace charge was partially melted in the recording period, 13:20-14:18.

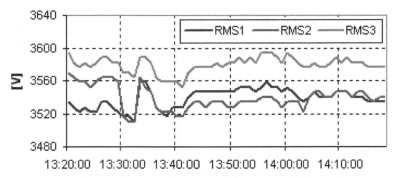

Fig. 22. RMS values of phase voltages in the intermediate state (MV Line)

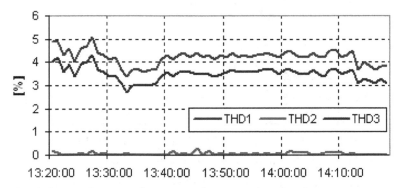

Fig. 23. THD of phase voltages in the intermediate state (MV Line)

In the intermediate state, THD of phase voltages do not exceed the compatibility limits, but are bigger comparatively with the cold state.

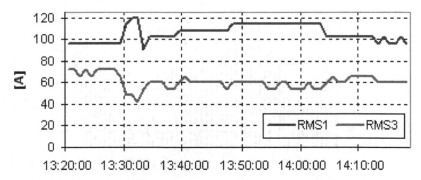

Fig. 24. RMS values of line currents in the intermediate state (MV Line)

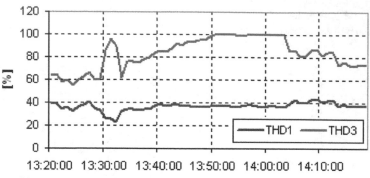

Fig. 25. THD of line currents in the intermediate state (MV Line)

In the intermediate state, the RMS values of the line currents show a poor balance between the phases. THD of line currents are remarkably high and exceed the compatibility limits.

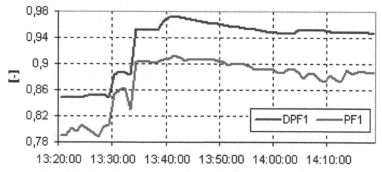

Fig. 26. DPF and PF per phase 1 in the intermediate state (MV Line)

The difference between the power factor and the displacement factor is significant in the intermediate state. This indicates the significant harmonic pollution and reactive power consumption.

PF per phase 1 is less than neutral value (0.92) almost all the time during the intermediate state. In the time period 13:20-13:35, PF is very small.

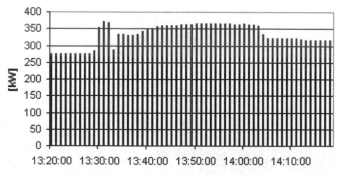

Fig. 27. Active power per phase 1 in the intermediate state (MV Line)

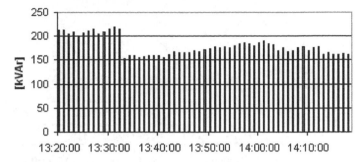

Fig. 28. Reactive power per phase 1 in the intermediate state (MV Line)

In the time period 13:20 - 13:35, the values of reactive power per phase 1 are almost equal to the values of active power. As a result, the power factor per phase 1 is very poor in the time period 13:20 - 13:35 (fig.26).

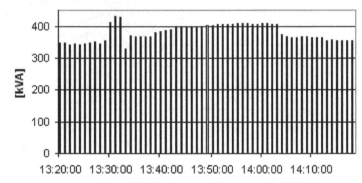

Fig. 29. Apparent power per phase 1 in the intermediate state (MV Line)

Fig.30-37 show the recorded parameters in the last stage of the heating. The furnace charge was totally melted in the recording period, 18:02-18:12.

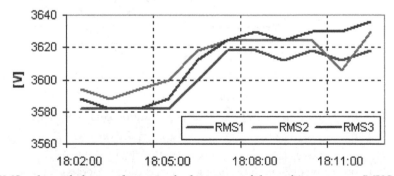

Fig. 30. RMS values of phase voltages in the last stage of the melting process (MV Line)

In the last stage of the melting process, THD of phase voltages are within compatibility limits, being smaller comparatively with the cold state or the intermediate state.

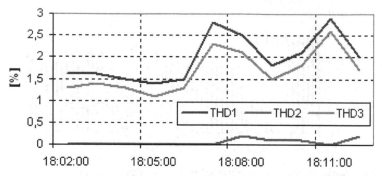

Fig. 31. THD of phase voltages in the last stage of the melting process (MV Line)

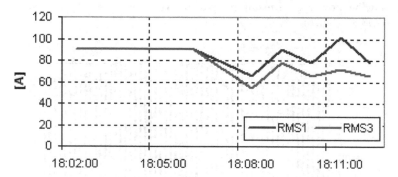

Fig. 32. RMS values of line currents in the last stage of the melting process (MV Line)

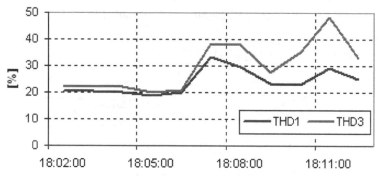

Fig. 33. THD of line currents in the last stage of the melting process (MV Line)

At the end of the melting process, the RMS values of line currents are much closer comparatively with cold state or intermediate state. THD of line currents exceed the compatibility limits, being of 20%...50% during this recording period.

The difference between the power factor and the displacement factor is small in the last stage of the melting process (fig.34). This indicates a decrease of harmonic disturbances and reactive power consumption (fig.36), comparatively with the cold state or the intermediate state.

In the time period 18:07 - 18:12, the values of reactive power per phase 1 increase; consequently, the power factor and the displacement factor per phase 1 decrease. Recorded values of active power per phase 1 are close to the apparent power values.

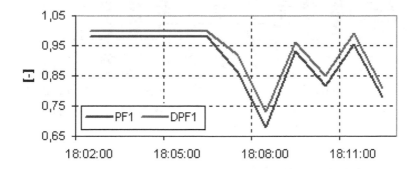

Fig. 34. DPF and PF per phase 1 in the last stage of the melting process (MV Line)

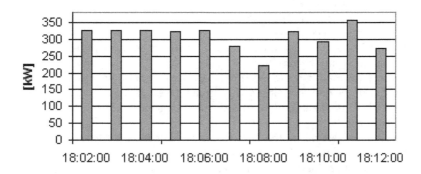

Fig. 35. Active power per phase 1 in the last stage of the melting process (MV Line)

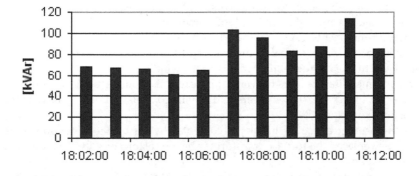

Fig. 36. Reactive power per phase 1 in the last stage of the melting process (MV Line)

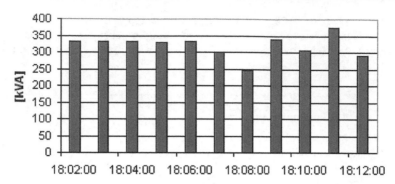

Fig. 37. Apparent power per phase 1 in the last stage of the melting process (MV Line)

## 6. Conclusion

The measurements results show that the operation of the analyzed furnace determines interharmonics and harmonics in the phase voltages and harmonics in the currents absorbed from the network.

THD of phase voltages are within compatibility limits, but voltage interharmonics exceed the compatibility limits in all the analyzed situations.

THD of line currents exceed the compatibility limits in all the heating stages. Because $I_{THD}$ exceed 30%, which indicates a significant harmonic distortion, the probable malfunction of system components would be very high.

THD of line currents are bigger in intermediate state comparatively with the cold state, or comparatively with the end of melting. This situation can be explained by the complex and strongly coupled phenomena (eddy currents, heat transfer, phase transitions) that occur in the intermediate state.

Harmonics can be generated by the interaction of magnetic field (caused by the inductor) and the circulating currents in the furnace charge.

Because the furnace transformer is in $\Delta/Y$ connection, the levels of the triple-N harmonics currents are much smaller on MV Line versus LV Line. These harmonics circulate in the winding of transformer and do not propagate onto the MV network.

On MV Line, 5th and 25th harmonics currents exceed the compatibility limits. The levels of these harmonics are higher on MV Line versus LV Line. Also, THD of line currents and THD of phase voltages are higher on MV Line versus LV Line, in all the analyzed situations.

The harmonic components cause increased eddy current losses in furnace transformer, because the losses are proportional to the square of the frequency. These losses can lead to early failure due to overheating and hot spots in the winding.

Shorter transformer lifetime can be very expensive. Equipment such as transformers is usually expected to last for 30 or 40 years and having to replace it in 7 to 10 years can have serious financial consequences.

To reduce the heating effects of harmonic currents created by the operation of analyzed furnace it must replaced the furnace transformer by a transformer with K-factor of an equal or higher value than 4.

Peak factors of line currents are high during the heating stages, and characterizes high transient overcurrents which, when detected by protection devices, can cause nuisance tripping.

The capacitors for power factor correction and the ones from Steinmetz circuit amplify in fact the harmonic problems.
PF is less than unity in all the analyzed situations. But, Steinmetz circuit is efficient only for unity PF, under sinusoidal conditions.
Under nonsinusoidal conditions, any attempt to achieve unity PF does not result in harmonic-free current. Similarly, compensation for current harmonics does not yield unity PF.
For optimizing the operation of analyzed induction furnace, it's imposing the simultaneous adoption of three technical measures: harmonics filtering, reactive power compensation and load balancing. That is the reason to introduce harmonic filters in the primary of furnace transformer to solve the power interface problems. In order to eliminate the unbalance, it is necessary to add another load balancing system in the connection point of the furnace to the power supply network.

# 7. References

Arrillaga, J., Watson, N. R., & Chen, S. (2000). *Power System Quality Assessment*, John Wiley and Sons, ISBN 978-0-471-98865-6, New York.

Ching-Tzong Su, Chen-Yi Lin, & Ji-Jen Wong (2008). Optimal Size and Location of Capacitors Placed on a Distribution System. *WSEAS Transactions on Power Systems*, Vol. 3, Issue 4, (april 2008), pp. 247-256, ISSN 1790-5060.

George, S., & Agarwal, V. (2008). Optimum Control of Selective and Total Harmonic Distortion in Current and Voltage Under Nonsinusoidal Conditions, *IEEE Transactions on Power Delivery*, Vol.23, Issue 2, (april 2008), pp. 937-944, ISSN 0885-8977.

De la Rosa, F. C. (2006). *Harmonics and Power Systems*, CRC Press, Taylor&Francis Group, ISBN 0-8493-30-16-5, New York.

Iagăr, A., Popa, G. N., & Sora I. (2009). Analysis of Electromagnetic Pollution Produced by Line Frequency Coreless Induction Furnaces, *WSEAS TRANSACTIONS on SYSTEMS*, Vol. 8, Issue 1, (january 2009), pp. 1-11, ISSN 1109-2777.

Lattarulo, F. (Ed(s).). (2007). *Electromagnetic Compatibility in Power System*, Elsevier Science&Technology Books, ISBN 978-0-08-045261-6.

Muzi, F. (2008). Real-time Voltage Control to Improve Automation and Quality in Power Distribution, *WSEAS Transactions on Circuit and Systems*, Vol. 7, Issue 4, (april 2008), pp. 173-183, ISSN 1109-2734.

Nuns, J., Foch, H., Metz, M. & Yang, X. (1993). Radiated and Conducted Interferences in Induction Heating Equipment: Characteristics and Remedies, *Proceedings of Fifth European Conference on Power Electronics and Applications*, Brighton, UK., Vol. 7, pp. 194-199, september 1993.

Panoiu, M., Panoiu, C., Osaci, M. & Muscalagiu, I. (2008). Simulation Result about Harmonics Filtering Using Measurement of Some Electrical Items in Electrical Installation on UHP EAF, *WSEAS Transactions on Circuit and Systems*, Vol. 7, Issue 1, (january 2008), pp. 22-31, ISSN 1109-2734.

Rudnev, V., Loveless, D., Cook, R., & Black, M. (2002). *Handbook of Induction Heating*, CRC Press, Taylor&Francis Group, ISBN 0824708482, New York.

Sekara, T. B., Mikulovic, J.C., & Djurisic, Z.R. (2008). Optimal Reactive Compensators in Power Systems Under Asymmetrical and Nonsinusoidal Conditions, *IEEE*

*Transactions on Power Delivery*, Vol. 23, Issue 2, (april 2008), pp. 974-984, ISSN 0885-8977.

CA8334, Three Phase Power Quality Analyser, technical handbook, Chauvin Arnoux, France, 2007.

IEC 61000-3-4, EMC, Part 3-4: Limits – Limitation of Emission of Harmonic Currents in Low-Voltage Power Supply Systems for Equipment with Rated Current Greater than 16A, 1998.

IEC/TR 61000-3-6, EMC, Part 3-6: Limits – Assessment of Harmonic Emission Limits for the Connection of Distorting Installations to MV, HV and EHV Power Systems (revision), 2005.

Power Quality for Induction Melting in Metal Production, TechCommentary Electric Power Research Institute (EPRI), U.S.A., 1999, available at:

http://www.energy.ca.gov/process/pubs/pq_inductn_melting_tc114625.pdf

# Harmonics Effect in Industrial and University Environments

M.H. Shwehdi

*King Faisal University, College of Engineering, Al Ahsa,*
*Saudi Arabia*

## 1. Introduction

### 1.1 Analyzing harmonic distortion produced from lead blast furnace (LBF)

A major cost to a steel factory facility is the energy used to power the arc furnace for the melting and refining process. Operation at low power factor results in additional voltage drop through the power system yielding a lower system voltage on the plant buses. Low system voltage increases the melt time and will add to the overall plant operating costs per ton. Low power factor can also result in additional costs in the form of penalties from the electric-utility company [1-2, 8]. Capacitor can be applied in steel factory facilities for a wide range of benefits. The capacitors will improve the power factor of the system; reduce billing penalties imposed by the electric power utility, and increase system voltage-boosting productivity. The system losses are also reduced improving the electrical system efficiency. However, harmonic sources in the steel mill can interact with capacitor banks resulting in problems if they are not properly applied. The effect of harmonics varies depending on the type of load. In some cases such as a resistance heating load all of the applied voltage does useful work; although, in most cases involving transformers and motors only the 60-Hz component of the voltage does useful work and the harmonic component generates useless heat. Sensitive electronic control circuits, timers, and logic circuits may be affected if the supply voltage is distorted [3-5].

The harmonic current generated by any non-linear load flows from the load into the power system. This current, seeking a low impedance path to ground, causes a voltage of the drop through the system according to Ohm's Law. The harmonic voltage combines with the 60 Hz voltages producing a distorted power system voltage. The harmonic laden power system voltage is then imposed on all of the remaining loads connected to the system this voltage distortion may result in more harmonic currents being produced as other linear loads experience the distorted system voltage.

A few industries like steel mills and aluminum smelters used electricity to power arc furnaces, which distorted the waveform, because the current flow was not directly proportional to the voltage. These loads are called non-linear loads. Non-linear loads cause waveforms that are multiples of the normal 60 Hertz sine wave to be superimposed on the base waveform. These multiples are called harmonics. Harmonic is defined as a sinusoidal component of a periodic wave having a frequency that is an integral multiple of the fundamental frequency. For example, the second harmonic is a 120 Hertz waveform (2 times 60 Hertz), the third is a 180 Hertz waveform, and so on.

Harmonic distortion may or may not create a problem for own facility. A plant may have harmonics present, but experience no adverse effects. However, as harmonic levels increase, the likelihood of experiencing problems also increases. Typical problems include:

- Malfunctioning of microprocessor-based equipment
- Overheating in neutral conductors, transformers, or induction motors
- Deterioration or failure of power factor correction capacitors
- Erratic operation of breakers and relays
- Pronounced magnetic fields near transformers and switchgear

To make matters worse, harmonics can sometimes be transmitted from one facility back through the utility's equipment to neighbouring businesses, especially if they share a common transformer. This means harmonics generated in one facility can stress utility equipment or cause problems in other neighbour's facility and vice versa.

Electric utilities have recognized this problem and are adopting standards, like the Institute of Electrical and Electronics Engineers (IEEE) Standard 519 which defines allowable harmonic distortion at customer service entrances. This standard is designed to protect both businesses and utilities, many other standards are also available and set limits for such harmonic penetration.

Harmonic analysis, load flow analysis, and power factor correction in Metal Scrab plant in Saudi Arabia, were considered for two reasons: 1) the planned installation of a new Induction furnace; and 2) the correction of the overall plant power factor to a value above 0.90 lagging to eliminate utility penalties.

## 2. The behavior of electric arc furnace

The voltage across an electric arc, which is relatively independent of current magnitude, consists of three components, anode drop, cathode drop and arc column component; which amount to about 12 volts/cm of arc length.

Typical values of arc voltages are in the range of 150-500 volts. Since the arc is extinguished at current zero, the power factor plays an important role on arc re-ignition. The figure 1. shows how arc voltage, power factor, input power; arc power and reactive power vary with arc current for a particular tap setting on the furnace transformer. The furnace is normally operated near maximum arc power, which corresponds to a power factor of 70% [9].

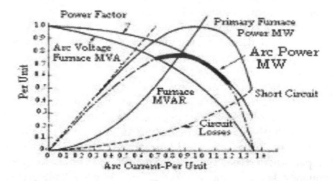

Fig. 1. Electrical Characteristics of Electric Arc Furnace

The three basic changes in operating states of an electric arc furnace, which can produce distinguishable voltage disturbances on power system, are open circuit condition, short circuit condition and the normal operation. The measurable data of interest for an electric arc furnace load include the following three phase quantities: supply voltage, real and reactive power, flicker, frequency and total harmonic distortion in respective phases. Because of the non-linear resistance, an arc furnace acts as a source of current harmonics of the second to seventh order, especially during the meltdown period. Voltage fluctuations are produced in this way through impedance on the value of harmonic currents supplied and the effective impedances at the harmonic frequencies. The harmonic current Iv of the arc furnace forms a parallel tuned circuit consisting of capacitor C with reactive power and mains inductance, resulting from the mains short circuit power. When this tuned circuit resonates at a harmonic frequency, its reactance is high and a harmonic voltage arises, which is damped by the resistance of the resistive component of the supply system consumers' equipment. The Q factor of this tuned circuit is low at times of full load, and no resonant peaks occur. But in slack periods with combinations of low load with high resistance and Q factor values, harmonic voltages are expected at levels sufficient to cause appreciable interference [11].

## 3. Harmonic mitigation

Several methods of mitigating harmonics have been developed over the years. The most common method is using filter, either passive or active. Passive filter block certain harmonic bandwidth while active filter injects current into the system to cancel the current harmonic waveforms. Both methods have their advantages and disadvantages, for example, advantage of passive filter is easy to design and active filter can monitor many frequencies simultaneously while disadvantage of passive filter is bulky in size and active filter is costly. Harmonic filters are useful and practical to be implemented by consumer near the proximity of the non-linear load at the low voltage system. Another method which is normally used by consumers is using phase cancellation method using twelve pulse converters instead of six pulse converters [12].

Similar application using filters for utility at higher voltage level such as distribution network requires extensive economic consideration. This is due to the size and cost of the equipment while most of harmonic pollutant is caused by consumer. There is little study on a feasible and cost effective means for utility to mitigate harmonic, especially harmonic voltage. A study was conducted on method using shunt harmonic impedance which can act like a central damper to reduce harmonic at distribution network [13]. This method is considered to be less expensive compared to active filter. The method uses power electronic to emulate resistive behavior for harmonic. However, the method is still under further study. Currently, all harmonic mitigation techniques involve equipment required to be installed on the system. There is yet a study on using other factors which can affects harmonic voltage distortion such as network impedance. Optimizing network impedance to mitigate harmonic can be cost effective for utility to apply. Because of mitigating harmonic is expensive, many utility company have resorted in imposing penalty to consumer for injecting current harmonic above the standard steady state limit into the system. This process requires method of determining harmonic contribution by the consumers  and the equipment need to be installed at all consumers' feeder which is very costly[13].

## 4. Brief steel plant system description

The steel plant system consists of 49 buses and 38 two winding transformers. The plant is fed from two utility substations at 230 KV and through four 230/34.5 KV transformers. From 34.5 KV many 34.5/13.8 KV transformers are installed to feed difference load including three electric Arc Furnaces (EAF1, EAF2, EAF3) and two Ladle Furnaces (LF1, LF2). Part of The single line diagram of the arc furnaces of this system is shown in figure 2.

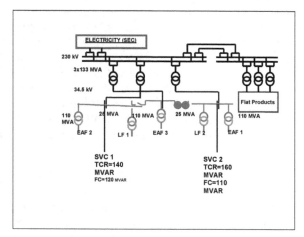

Fig. 2. Partial single line diagram of steel plant system.

This system was simulated by using a software package and the results of load flow, total harmonic distortion, and power factor at some buses are shown in table 1.

| Bus# | Nominal Voltage (KV) | LF Voltage (p.u) | LF Angle (deg) | THD (%) | Power Factor |
|------|---------------------|------------------|----------------|---------|--------------|
| 1 | 230.00 | 1.00 | 0.00 | 6.47 | 56 |
| 2 | 34.50 | 0.95 | -1.90 | 10.31 | 59 |
| 3 | 34.50 | 0.95 | -1.90 | 10.31 | 59 |
| 5 | 34.50 | 0.95 | -1.90 | 10.30 | 59 |
| 45 | 0.48 | 0.91 | -3.80 | 10.11 | 58 |
| 46 | 0.48 | 0.91 | -3.80 | 10.11 | 58 |

Table 1. Load flow, THD, and power factor results

| Filter Location | Order | Rated KV | Kvar | XL | Q |
|-----------------|-------|----------|------|-----|---|
| Bus # 1 | 2nd | 230.00 | 35035.00 | 377.47 | 75.00 |
| Bus # 5 | 5th | 34.50 | 30572.00 | 79.00 | 39.50 |
| Bus # 5 | 7th | 34.50 | 24745.00 | 120.20 | 84.10 |

Table 2. Filtre Data

| Bus # | Nominal Voltage (KV) | LF Voltage (p.u) | LF Angle (deg) | THD (%) | Pf |
|-------|----------------------|-------------------|-----------------|---------|-----|
| 1 | 230.00 | 1.00 | 0.00 | 1.24 | 0.98 |
| 2 | 34.50 | 0.99 | -2.30 | 2.95 | 0.99 |
| 3 | 34.50 | 0.99 | -2.30 | 2.95 | 0.99 |
| 5 | 34.50 | 0.99 | -2.30 | 2.95 | 0.99 |
| 45 | 0.48 | 0.95 | -4.00 | 2.92 | 0.97 |
| 46 | 0.48 | 0.95 | -4.10 | 2.92 | 0.97 |

Table 3. Load Flow, THD, and Power Factor Results

Single-tuned filters were designed for the metal plant system according to the next paragraph theories and its input data are provided in table 2. The results of load flow, total harmonic distortion, and power factor of buses 1, 2, 3, 5, 45, and 46 after installing filter are shown in table 3. Also the spectrum and waveform of bus 34.5 kV is provided in figure 3.

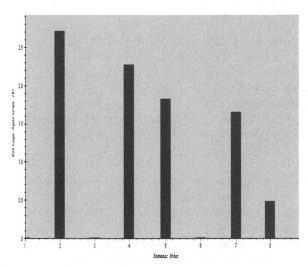

Fig. 3. Spectrum of voltage at bus 34.5 kV

## 5. Filter design

Harmonic filters are designed to suppress system harmonics as well as to improve power factor. They allow a system to meet IEEE Standard 519 harmonic limits while avoiding power factor penalties. Filter designs are tailored to individual project objectives such as meeting a harmonic limit and/or a power factor level. This is a complex and involved engineering task, where alternative designs are checked to ensure that the final one will meet study objectives [14].

Typically, single tuned shunt filters are designed with a reactor and capacitor in series, connected as a shunt load on the system. In more complex studies, other types of filters, like 2nd or 3rd order, C-type, and/or double tuned filters are designed. In a complex filter design, a combination of filter types could be required. A generic term used to describe those types of equipment whose purpose is to reduce the harmonic current or voltage flowing in or being impressed upon specific parts of an electrical power system, or both [4,11].

The filter is tuned slightly below the harmonic frequency of concern. This allows for tolerances in the filter components and prevents the filter from acting as a direct short circuit for the offending harmonic current. Further allows the filter to perform its function while helping to reduce the duty on the filter components. It also minimizes the possibility of dangerous harmonic resonance should the system parameters change and cause the tuning frequency to shift slightly higher [6, 7].

Once the filter type and the components (reactors, capacitors and resistors) are determined, the design program is used to model these filters. Overall power system operation can then be analyzed to determine the effectiveness of the filtering scheme. Ratings of all filter components along with the protection schemes and control methods are identified, and detailed specifications are developed for the manufacture of the filters. All designs are based on relevant IEEE Standards for capacitors and reactors [15].

## 5.1 Filtre components
### 5.1.1 Capacitors
Capacitors are composed of standard units that are connected in series or parallel for obtain the desired overall voltage and KV rating [5]. The capacitor's are designed and chosen with the following considerations:
- Harmonic current peaks have a 100% coincidence.
- Nominal System over-voltage of 5%.
- Ambient voltage distortion equal to the limits set forth by IEEE 519.
- Adherence to IEEE/ANSI peak and RMS voltage ratings.

### 5.1.2 Inductors
Inductors used in filter circuit need to be designed bearing in mind the high frequencies involved. Inductors rating depend mainly on the maximum RMS, current. The inductors and resistors form the ground side of a tuned filter [5]. The reactor current ratings are based on the following considerations:
- The reactor core will not saturate for currents less than
  250% of the fundamental current rating of the filter bank.
- Peak flux density of the core will be less than 1.2 – 1.4
  Tesla assuming all harmonic current peaks is 100% coincident.

## 5.2 Tuned filter
A single tuned filter is a series RLC circuit tuned to the frequency of one harmonic .its impedance is given by

$$Z_1 = R + j(wL - 1/wC) \qquad (5.1)$$

Which at the resonant frequency $f_n$ reduces to R. There are two basic design parameters to be considered prior to the selection of R, L and C. these are the quality factor Q, and the relative frequency deviations. It is generally more convenient to deal with admittances rather than impedance in filter design

$$Y_f=1/R(1+j2s\ Q) =G_f + j\ B_f \tag{5.2}$$

Where

$$G_f=Q/X0(1+4s^2Q^2) \tag{5.3}$$

$$B_f=2sQ^2/X0(1+4s^2Q^2) \tag{5.4}$$

$$X_0= \sqrt{\frac{L}{C}} \tag{5.5}$$

The harmonic voltage at the filter bus bar is

$$V=I/Y_f+Y_s \tag{5.6}$$

Therefore, to minimize the voltage distortion it is necessary to increase the overall admittance of the filter in the parallel with the a.c system. The harmonic voltage increases with (s) [4]. In term of Q and s can be equation (6) can be written as follows:

$$V=I\{(Gs+1/R(1+4s^2Q^2))^2+(Bs-2sQ/R(1+4s^2Q^2))^2\}^{-1/2,} \tag{5.7}$$

## 6. University Personnel Computers (PC) effect on line currents harmonics

### 6.1 Introduction

Power Quality problems are increasing with the proliferation of nonlinear devices, which draw none sinusoidal current waveforms when supplied by a sinusoidal voltage source. When these devices are present in an electric power system, they cause harmonic distortion of voltages and currents. Individually, single phase nonlinear load may not pose many serious harmonic problem, but large concentrations of these loads have the potential to raise harmonic voltages and currents to unacceptable high levels which results in increased neutral currents in four wire system, over heating of distribution system components and may cause mechanical oscillations in generators and motors. Other unwanted effects are capacitor and insulation failure due to harmonic resonance, malfunction of installed protection systems, transient voltage fluctuations, over heating of system transformer and cables, error of power electronic equipments operations and telephone interference.

Many desktop personal computers still present a nonlinear load to the AC supply. This is because they have a power supply design known as a "capacitor input switch mode power supply". Much of today's Information Technology equipment including servers, routers, hubs, and storage systems almost universally use a different power supply design known as "Power Factor Corrected". These devices present a very linear load to the AC supply and do not generate harmonic currents. In fact they are one of the cleanest loads on the power grid and generate less harmonic current than many other devices such as fluorescent lighting or variable speed drives. The 3rd harmonic currents, the predominant harmonic in PC power supplies, causes overloaded neutrals, overheated transformers, and annoyance circuit breaker tripping. Very high price may be spent on equipment which will either filter or block the harmonics or withstand the heating effects of the harmonics.

Studies on the monitoring of power quality at computer sites have been conducted as early as 1969, [16, -18], and continuing interest in this area has maintained regular publications thereafter [19-21]. Most early studies were concerned with the effects of power disturbances on the correct operation of the computer facility. However, with the advent of relatively inexpensive personal computers (PC), the emphasis of computer power quality monitoring has also moved towards investigating the effects that large concentrations of PCs can have on other utility customers.

Personal computing impacts on power quality are increasing due to the common place usage of switched mode power supplies (SMPS) for converting single phase AC into low voltage DC for supplying processing electronics. Such power supplies, which are responsible for the generation of odd line current harmonics, are the main concern of this paper. However, in turn harmonically polluted line currents can distort supply voltages causing power quality problems for other consumers connected at a point of common coupling [22]. Additionally, and somewhat ironically, the switched mode power supply itself can be affected by non-sinusoidal supply voltages [23], which can increase or decrease current harmonics depending on the nature of the voltage distortion.

Switched mode power supplies are by no means restricted to PCs and can be found in a variety of other widely used electrical equipment including low energy lighting, battery chargers, televisions and their peripherals.

A recent study [24] has shown that the line current harmonics from a single PC differed considerably to the harmonics generated collectively by several PCs of the same type. One widely held theory [25] regarding this effect introduces the concepts of *attenuation* and *diversity*. Attenuation describes the reduction in harmonic magnitude, and change in phase angle, as a load connected to a SMPS increases, and attributes this effect to the change in the spectrum of the line current pulse which widens to allow more power flow through the SMPS. Attenuation is also observed where several identical loads share the same source impedance. Diversity describes a The influence of personal computer processing modes on line current harmonics similar effect where a reduction, or even cancellation, of harmonics is possible due to loads of different levels, or connected through different impedances, presenting differing phase angles to the supply. These findings have not been proved using large-scale studies; although predictions based on results taken from individual computers have been reported [26].

The primary aim of this investigation is to investigate how the mode of operation of a PC affects the harmonics produced in the line current. This is an area barely mentioned in previous literature although these effects are closely related. One published study has made limited investigations of this type, but, again, only for individual computers [27]. In new construction or renovation, many power disturbances can be prevented or significantly lessened by designing for power quality assurance, at surprisingly small cost.

In view of the concerns regarding cumulative effects of large collections of PCs, this study was conducted within a University library building containing over 370 PCs. Furthermore, the study was intended to investigate the primary effect on line current harmonics caused by mode of operation, in isolation from additional secondary effects caused by distorted supply voltages. Investigation of this primary effect was achieved by monitoring during periods when the PCs represented the only load on the transformer supplying the library building and consequently the supply voltage waveforms were relatively undistorted. Most of these disturbances originate right within the building. Personal computers, laser printers and other switched-mode power supply equipment within your building are usually the culprits for most of the power supply irregularities affecting other computers. It's a problem that has only recently begun to be recognized in the building industry, as more and more computers and similar equipment are installed, turning the modern office or factory into a high-tech computer environment. [28]

The objective of section six is to investigate the effect of an University-PC and other harmonic contributing sources on the line currents that may have influence on processing modes and to measure the actual Total Harmonic Distortion during such modes.

## 7. Power quality issues related to computers

Although harmonics is the subject of several international standards [16], the difficulties of regulating harmonics emissions from small, individual devices employing switched mode technology, such as a personal computer, is compounded by the fact that the harmonic generation capabilities of these devices varies according to the number of parallel connected devices.

In the last decade, power quality has become a major issue for electric utilities and their customers, and indirectly to almost all manufacturers of equipment that depend on sinusoidal supply voltage waveforms. The proliferation of solid-state switching devices in both industry and home has increased the harmonic levels found on electricity supply systems. In the past, sources of harmonics were limited and in most cases tolerable. However today, virtually all high performance electronic equipment uses switched mode power converters, conducting current for part of a cycle through multiple paths in order to efficiently convert alternating current to direct current. In response to this increased demand some countries encouraged the development of more efficient ways of getting full benefits from electrical energy by various switching methods.

On the other hand a new technique of manufacturing circuits called integrated electronic circuits have been invented and led directly to increasingly complex systems such as in industry ,communications, domestic appliances ... etc   because its size, weight and cost. These circuits, although increased speed of operations and complexity of tasks, typically use a technique called switch mode power supplies [26- 29].

The majority of modern electronic units use switched mode power supplies (SMPS). The advantage–to the equipment manufacturer– is the size, cost and weight is significantly reduced and the power unit can be made in almost any required factor [28]. But unfortunately switch mode power supplies act as non–liner loads or they can be considered as a noise generators which get down the quality of the electricity supply.

Equipments with switch mode circuits are usually more sensitive to main variations and disturbances than liner loads [30]. For this reason power quality has become a major issue for electric utilities and their costumers and almost all manufactures of equipment that despond on sinusoidal supply voltage waveforms. The increasing of solid state switch devices such as Personal computers in both industry and homes will draw rather than continuous current from the supply; the power supply unit will draw pluses of current which contain large amount of harmonics [29, 32].

The switching mode power supplies used in personal computers are major sources of harmonic currents. An experiment [26, 28, and 32] has been done for different types of computers to measure the harmonic generated by each type and the result was scheduled as:

|       | Mac IIsi | IBM XT | IBM PS/30 | Mac Plus | IBM PS/70 |
|-------|----------|--------|-----------|----------|-----------|
| $I_1$ | 106.82 (100%) | 165.93 (100%) | 199.02 (100%) | 83.35 (100%) | 541.88 (100%) |
| $I_3$ | 89.31 (83.6%). | 122.45 (73.8%) | 152.63 (76.7%) | 72.30 (86.7%) | 458.83 (84.7%) |
| $I_5$ | 65.18 (61.0%) | 68.03 (41.0%) | 94.76 (47.6%) | 55.80 (66.9%) | 333.54 (61.5%) |
| $I_7$ | 40.51 (37.9%) | 26.71 (16.1%) | 39.16 (19.7%) | 36.53 (43.8%) | 192.04 (35.4%) |
| $I_9$ | 21.25 (19.9%) | 5.64 (3.4%) | 10.39 (5.2%) | 17.78 (21.3%) | 71.18 (13.1%) |
| $I_{11}$ | 8.70 (8.1%) | 3.98 (2.4%) | 3.53 (1.8%) | 3.31 (4.0%) | 17.75 (3.3%) |
| $I_{13}$ | 2.95 (2.8%) | 8.46 (5.1%) | 1.49 (0.7%) | 4.95 (5.9%) | 49.02 (9.0%) |
| $I_{15}$ | 3.46 (3.2%) | 8.29 (5.0%) | 2.30 (1.1%) | 7.51 (9.0%) | 48.01 (8.8%) |

Table 4. Magnitudes of harmonic currents of various PC's in mA [32]

The phase current is given by:

$$I_P = \sqrt{I_1^2 + I_3^2 + I_5^2 + I_7^2 + I_9^2 + I_{11}^2 + I_{13}^2 + I_{15}^2} \tag{6.1}$$

The IBM XT™ has the lowest percent third harmonic current (74%). Since the computer is connected line-to-neutral in a 3-phase system, the neutral current is approximately equal to three times the vector sum of the third and ninth harmonic currents flowing in each phase.

$$I_n = 3\sqrt{I_3^2 + I_9^2} = 367,764 mA \tag{6.2}$$

$$I_P = 219,226 mA \tag{6.3}$$

$$\frac{I_n}{I_P} = 1,677 \tag{6.4}$$

The Mac Plus™ offers the worst case third harmonic current (87%). In a similar manner $I_n$ & $I_p$ are obtained.

$$I_n = 223,377 \tag{6.5}$$

$$I_P = 130,507 mA \tag{6.6}$$

$$\frac{I_n}{I_p} = 1,712 \tag{6.7}$$

With a large number of personal computer loads, the neutral current is expected to be 1.7 times the phase current. It will certainly overload the neutral conductor that is designed to handle lower currents than the phase currents [14]. Non linear loads connected to a sinusoidal source inject harmonic currents into the source. Much attention has been focused on large power converters as sources of harmonics due to the high magnitude of the currents evolved. Nevertheless, harmonics generated by low power converters become significant when large numbers of converters are used simultaneously [14, 15]. Indubitably the personal computer is an ideal place for these converters. The power supplies employed in most personal computers are of the switching mode type. In switching mode power supply, the 60 Hz AC voltage is converted to into DC through a single phase rectifier and the output voltage of the rectifier is stepped down using AC to DC converter [13, 14]. The main advantage of the switching mode power supply over the traditional linear power supply is its high energy efficiency since the switching elements used (BJT's, MOSFETS) are either completely off or completely on.

Personal computers impacts on power quality are noticeable due to the common place usage of switched mode power supplies (SMPS) for supplying processing electronics. Such power supplies are responsible for the generation of odd line current harmonics. However, in turn harmonically polluted line currents can distort supply voltages causing power quality problems for other consumers connected at a point of common coupling (PCC). Switched mode power supplies are by no means restricted to PCs and can be found in a variety of other widely used electrical equipment including low energy lighting, battery chargers, televisions and their peripherals. For power quality considerations, the PC is the most significant item for a variety of reasons including.

- On an individual basis, it represents one of the larger loads consuming typically 200-300 W per unit.
- The current drawn is affected by user behavior.
- Large numbers of PCs can be connected to a single supply point.

The last point is one of the major concerns of increasing PC usage and has resulted in power quality studies of the cumulative effects of many PCs connected to the same bus bar [27]. A recent study has shown that the line current harmonics from a single PC differed considerably to the harmonics generated collectively by several PCs of the same type. One widely held theory regarding this effect introduces the concepts of attenuation and diversity [25].

Attenuation describes the reduction in harmonic magnitude, and change in phase angle, as load connected to a SMPS increases due to the change in spectrum of the line current pulse which widens to allow more power flow through the SMPS. Attenuation is also observed where several identical loads share the same impedance. Diversity describes a similar effect where a reduction or even cancellation of harmonics is possible due to loads of different levels, or connected through different impedances, presenting differing phase angles to the supply. These findings have not been proved using large scale studies [25-28].

## 8. Study proceedures and results

### 8.1 Introduction

The University consists consists of 60 buildings including students housing. Almost each student at the university has a laptop regardless of the number of PC's in each, laboratories, or rooms. The university in-house that there are around 6000 PCs distributed unequally . Most of the PS's in the university are of model HP, DELL, Compaq dc 7800P; they use Intel core 2 Dual processors.

### 8.2 Measurement plan

This investigation adapted the following steps which include:
- Obtaining all line diagrams of the area of investigation.

- Securing the total number of the university PC's and most PC's concentration area.
- Circulate Questionnaire to major PC users such as ITC to develop a sense where are the major area that may have harmonics as to affect ITC line currents and servers etc...
- Locate and prepare the building (14, consist of 263 PC's and 58 consist of 280 PC's) switchboard to be able to conduct different measurements being the highest building containing PC's.
- Conduct Harmonic measurements at these buildings and monitor and check harmonics at different loading processing modes and times.
- Recommendation and findings are to be clearly drawn out of results and stressing the mode of operations and size of PC lab.
- Identify the sources and causes of harmonics at such selected locations: use such data in the process of making the appropriate mitigating actions.

To conduct harmonic measurement the team members attended intensive lectures and training on Power Quality and measurements. Training focused on how to use the power quality analyzer (PQA).

Single line diagrams for each of the building under investigate were obtained as shown in Figure 4. Permissions to conduct measurement and open switch boards of the different feeders were also obtained through the university Electrical Maintenance department.

The whole Excl file containing total PC's provided by from ITC indicating all the university PC's record and their distributions at the different building and labs etc...The total February 2009 PC's numbers was 6,344.00.

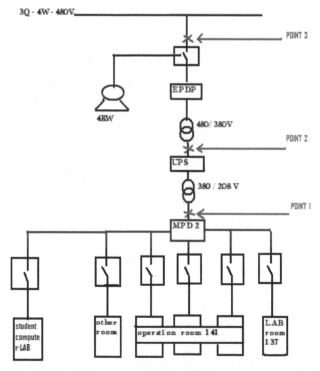

Fig. 4. Building 14 Single Line Diagram and Points of Measurements and switch boards

## 8.3 Measurements at normal load

The measurements have been conducted by the team members of the project according to the following time Table:

| Day | Reading #1 period | Reading #2 period | Reading #3 period | location inside the substation |
|---|---|---|---|---|
| Monday | 9:00AM-12:00Pm | 12:10Pm-1:30PM | 1:40PM-4:00PM | At point 1* |
| Tuesday | 9:00AM-12:00Pm | 12:10Pm-1:30PM | 1:40PM-4:00PM | At point 2** |
| Wednesday | 9:00AM-12:00Pm | 12:10Pm-1:30PM | 1:40PM-4:00PM | At point 3*** |

*indicates the point between MPD2 &the transformer 380/208 v
**indicates the point between UPS &the transformer 480/380 v
***indicates the point between EPDP & the main feeder 3ф-4w 480v

Table 5. Time table of measurements

At the first point shown in the single line diagram (Figure 4), three measurements have been conducted over the three periods as listed in table 5.

## 8.4 Measurements and results

The measurement of the harmonics caused by PC's and any other nonlinear elements were conducted at each bus (points 1, 2, and 3) as indicated in the single line diagram.

At the first point shown in the single line diagram (Figure 4), three measurements have been conducted over three periods. The THD during the first period at point one was measured and the maximum total harmonic distortion (THD) was 10.6%. The THD during the second period at point one and the maximum total harmonic distortion (THD) was 10.7%. The THD during the third period at point one and the maximum total harmonic distortion (THD) is 10.8%

From the previous three readings it is clear that at normal load at point one the maximum total harmonic distortion of the three readings is 10.8%. Moreover, the harmonics effect appears in the current waveform as in Figure 5.

For the second point the THD during the first period indicated a maximum total harmonic distortion (THD) was 29%.

On the second measuring period the THD during the second period at point one and the maximum total harmonic distortion (THD) was 28.4%. The THD during the third time period at point two indicated maximum total harmonic distortion (THD) of 28.9%. From the previous three readings it is clear that at normal load at point two increases the harmonic flow, the maximum total harmonic distortion of the three readings is 28.9%. Moreover, the harmonics appears to distort the phase current waveform as in Figure 6.

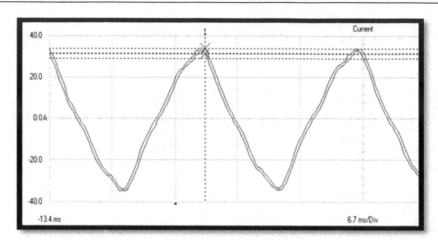

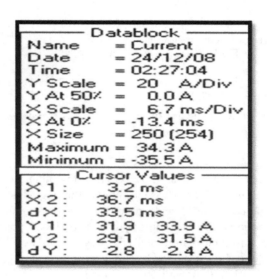

Fig. 5. Current wave forms at point one

The sum of all these THD's = 83.8% is not equal the THD at point one which is 10.8%.The difference between these two values explains which is called harmonic cancellations. The phase angles of the magnetization current harmonics oppose the phase angles of the load current harmonics and such lead to harmonic cancellations. The maximum total harmonic distortion at point three is 20.6% which is less than that at point two (29%).The difference between the two values is caused also by harmonics cancellation. According to the instructions provided with the power quality analyzer manual and other standards which state that if the current THD is less than 20% the harmonic distortion is probably acceptable, the total harmonic distortion at point three (29%) was greater than 20%, so it is not acceptable and definitely will make adverse affect on the neutral line cable.

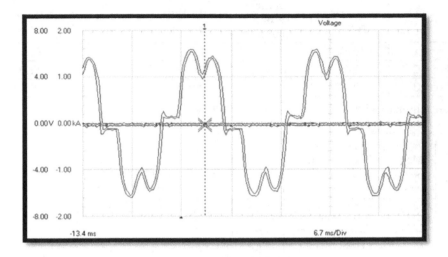

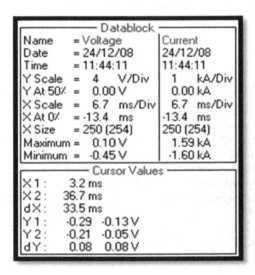

Fig. 6. Current wave forms at point two

To avoid the injection of harmonics into the system, a harmonic filter must be designed to eliminate the highest harmonics order flowing and then installed at the near bus of such produced harmonic. It is noticed in table 6 that the highest THD measured was during the period where students were using laboratories of Building 14 at point 2 with the highest harmonic is of fifth order. While table 7 illustrates harmonics produced as to the total numbers of Pc's used.

| Location | Max THD in the first period | Max THD in the second period | Max THD in the third period | Crest Factor | Highest Harmonic Order |
|----------|-----------------------------|------------------------------|-----------------------------|--------------|------------------------|
| Point 1  | 10.6% | 10.7% | 10.8% | 1.5-1.6 | Third |
| Point 2  | 29% | 28.4% | 28.9% | 1.5 | Fifth |
| Point 3  | 19.6% | 19.9% | 20.6% | 1.5 | Fifth |

Table 6. Summary of the obtained results for the three points of Bldg. 14

| No. of PC's | %age Mag. of 3rd Harmonic | %age Mag. of 5th Harmonic | %age Mag. of 7th Harmonic | %age Mag. of 9th Harmonic | %age Mag. of 11th Harmonic | %age Mag. of 13th Harmonic | %age Mag. of 15th Harmonic | %age Mag. of 17th Harmonic | THD % |
|------|------|------|------|------|------|------|------|------|------|
| 263 | 28.3A | 14.8A | 5.2A | 0.4A | 0.7A | 0.1A | 0.4A | 0.1A | 10.4 |
| 204 | 27.1A | 13.1A | 4.6A | 1.4A | 0.7A | 0.4A | 0.7A | 0.1A | 9.9 |
| 170 | 25.6A | 12.5A | 4.2A | 1.8A | 0.8A | 0.4A | 0.5A | 0.1A | 9.6 |

Table 7. Harmonic magnitudes for different No. of PC's at point 1

## 8.5 Standards and limits

International Standards have set some limits to the most equipment on the permissible harmonic content in the electrical system during operation. The following are general limits for various electrical equipments:

a.    Synchronous machine: permissible stator current distortion < 1.4%
b.    Asynchronous machines: permissible stator current distortion; 1.5% to 3.5%
c.    Cable: permissible core-shielding voltage distortion < 10%
d.    Electronic equipment: 5% voltage distortion with a maximum individual percentage of 3% depending on the equipment
e.    Transformer: permissible current distortion <5% at full load [ IEEE-519 ]

Most utilities have adopted standards to limit the harmonic content at the point of common coupling (PCC). Some of the Standards adapted around the world include:

• US/Canada IEEE 519
• Europe IEC61000 3-2, 3-4
• United Kingdom G5/4 -1
• China GB/T 14549

## 9. Thd theory and verification calaculations

The percentage of Total Harmonic Distortion (%THD) can be defined in two different ways, as a percentage of the fundamental component (the IEEE definition of THD) or as a percentage of the RMS (used by the Canadian Standards Association and the IEC).

$$\text{THD} = \frac{\sqrt{\sum_{n=2}^{\infty} I_{rms,n}^{2}}}{I_1} \tag{9.1}$$

Where, $I_{rms, n}$ is the amplitude of the harmonic component of order n (i.e., the nth harmonic). The numerator gives the RMS current due to all harmonics and $I_1$ is the RMS value of fundamental component of current only. Given above is the mathematical form of the IEEE. According to IEC standards, the mathematical form of THD is given below:

$$THD = \frac{\sqrt{\sum_{n=2}^{\infty} I_{rms,n}^2}}{I_1} \qquad (9.2)$$

Where

$$I_{rms} = \sqrt{\sum_{n=1}^{\infty} I_{rms}^2} \qquad (9.3)$$

Where $I_{rms, n}$ is the amplitude of the harmonic component of order n (i.e., the nth harmonic) and $I_{rms}$ is the rms value of all the harmonics plus the fundamental component of the current. The later standard is referred in this study, because the apparatus used for analysis was based on IEC Standards.

The 3rd, 5th, 7th and 9th harmonics being the most significant, the definition of THD may be modified and written as in 9.4

$$THD = \frac{\sqrt{I_{rms,3}^2 + I_{rms,5}^2 + I_{rms,7}^2 + I_{rms,9}^2}}{I_{rms}} \qquad (9.4)$$

The value of THD may be calculated for any number of computers using formula (9.3).
$I_{rms} = 308.4$ A
RMS magnitude of 3rd Harmonic= 25.6 A
RMS magnitude of 5th Harmonic= 12.5 A
RMS magnitude of 7th Harmonic= 4.2 A
RMS magnitude of 9th Harmonic= 1.8 A

$$THD = \frac{\sqrt{25,6^2 + 12,5^2 + 4,2^2 + 1,8^2}}{308,4} = 9.36\% \qquad (9.5)$$

Figure 9 is showing the magnitude of individual harmonics, when 263 PCs in building 14 were connected to the supply mains.

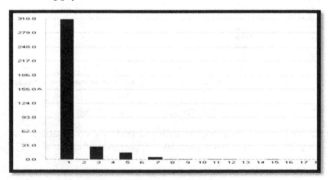

Fig. 7. Harmonic spectrums at point one when 263 PC's operating

The online value of THD was 9.6%. The percentage difference (Error) of the calculated and experimental value is 0.24%.

This difference caused by neglecting other odd harmonics such small error proves the validity of measurement using this PQA and it consequently plays a pivotal role for the accurate analysis of the odd harmonics.

Figure 8 explains the overall impact of individual harmonics cumulatively. Total Harmonic Distortion (THD) in line current is increasing when the electronic loads are increasing. Among odd harmonics only third harmonic plays active role whereas the other odd harmonics impact with increase in electronic loads is negligible.

By using linear interpolation, the relation between THD in current and the number of PC's (N) is given by the following equation:

$$I_t = 8.143 + 0.008 \times N \qquad (9.6)$$

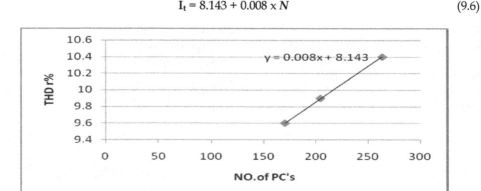

Fig. 8. THD curve vs. No. of PC's at point 1.

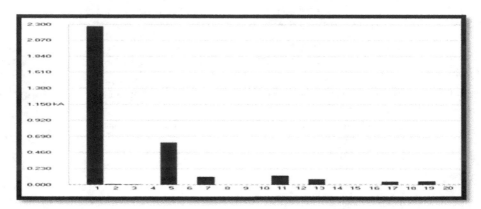

Fig. 9. Harmonic spectrums at point 2 when 263 PC's operating

The value of THD may be calculated for any number of computers with formula (9.4). Figure 9 is showing the magnitude of individual harmonics, when 263 PCs were connected to the supply mains.

$I_{rms} = 2271$ A

RMS magnitude of 3rd Harmonic= 3 A

RMS magnitude of 5th Harmonic= 600 A
RMS magnitude of 7th Harmonic= 112 A
RMS magnitude of 9th Harmonic= 1A

$$\text{THD} = \frac{\sqrt{3^2 + 600^2 + 112^2 + 1^2}}{2271} = 26.7\% \tag{9.7}$$

The online value of THD was 26.8%. The percentage difference (Error) of the calculated and experimental value is 0.1%.

## 10. Thd measurements discussion

According to the previous measurements it has been observed that the total harmonic distortion at point two (29 %) is much greater than that at point one (10.8%). Since there is no load connected between these two points except the Uninterruptible power supply (UPS), it is considered that UPS is the main reason for this difference. The UPS can be considered to fit 'in-line' between the loads and the mains power supply. In addition to providing power protection to the loads, it should also protect the main power supply itself from getting any harmonics generated by the loads themselves. However, it is again not commonly known that UPS and their design being power electronics oriented, also generate harmonic pollution. For any UPS this is typically stated as Total Harmonic Distortion (THD). Care has to be taken when comparing different THD values as these can differ when contrasting the two different types of on-line UPS (transformer-based and transformer less) and also with regard to the percentage of load applied for each measurement.

| No. of PC's (N) | %age Mag. of 3th Harmonic | %age Mag. of 5th Harmonic | %age Mag. of 7th Harmonic | %age Mag. of 9th Harmonic | %age Mag. of 11th Harmonic | %age Mag. of 13th Harmonic | %age Mag. of 15th Harmonic | THD r% |
|---|---|---|---|---|---|---|---|---|
| 263 | 1.6 | 33.7 | 6.4 | 0.1 | 7.6 | 4.4 | 0.1 | 19.7 |
| 204 | 1.6 | 33.6 | 4.7 | 0.1 | 7.9 | 3.5 | 0.1 | 20.6 |
| 170 | 1.4 | 33.5 | 4.3 | 0.1 | 7.8 | 3.5 | 0.1 | 20.6 |

Table 8. Magnitudes of harmonics for different numbers of PC's at point 3

Table 8 indicates the online value of THD is 19.7%. The difference of the calculated and experimental value of 0.37% as shown in table 9. This difference caused again by other odd harmonics being neglected, however, such low error proves the validity of measurement and it consequently plays a pivotal role for the accurate analysis of the odd harmonics.

| Location | Calculated values | Experimental values | %age Error |
|---|---|---|---|
| Point 1 | 9.36 | 9.60 | 0.24 |
| Point 2 | 26.7 | 26.8 | 0.10 |
| Point 3 | 19.3 | 19.7 | 0.37 |

Table 9. Comparison of calculated and experimental values when 263 PC'S were connected

Within a UPS it is rectifier that connects to the mains power supply and converts the mains alternating current (AC) into the levels of direct current (DC) required to power the inverter and charge the battery.

For transformer-based UPS, rectifiers are typically six or twelve-pulse, dependent upon the thyristor number and configuration. A six-pulse rectifier at full load will typically generate a THD of around 29% and a 12-pulse around 8%. To reduce these values further a passive harmonic filter can be installed alongside the UPS. The obvious disadvantages of this approach being increased capital cost, wiring, installation, loss of efficiency and increased footprint. Harmonic filters can be added post-installation but further installation costs and downtime need to be planned for.

The maximum total harmonic distortion at point three is 20.6% which is less than that at point two (29%).The difference between the two values is caused also by harmonics cancellation.

## 11. Conclusions

The nature of such metal factories are to expand because of the high and rapid demand on steel, aluminum, etc... to coup up with the higher rates of development. As for the plant and due to presence of three arc furnaces and two ladle furnaces and adding 1 Induction Furnace in this metal facility, one expects harmonics are considerably high in the steel plant without any filtering. Also, due highly inductive load of this steel plant the Power factor needs to be corrected to match that of the utility [8].

Harmonic measurements and analysis have been conducted and are becoming an important component of the plant routine measurements and for power system planning and design. Metal plant engineers are striving to meet with utility, and IEEE standard for harmonics as well as power factor. Considerable efforts have been made by the plant engineers in recent years to improve the management of harmonic distortion in power systems and meet the utility requested power factor levels.

Results obtained from steel plant system the power factor are low at about seven buses one of them bus number 1 the utility bus were the power factor found 0.56. The power factor of all the buses ranged between 0.56 and 0.59 which considered very low for the utility power factor which is 0.93. Results obtained from the harmonic studies indicate again that many buses of the plant including the utility bus have violated the IEEE-519 1992 standard. One has to remember that using software to analyze the practical conditions it is important to understand the assumption made and the modeling capabilities, of the non-linear elements.

The authors have met with plant engineers and discuss mitigation of the harmonic level as well as improvement of the power factor. Harmonic filters were designed to suppress low harmonic order frequencies and were installed at the different buses, the filtered harmonic of this plant were mainly for the 2nd, 5th, and 7th harmonics.

The plant operations with installation of the designed filters have improved the power factors to reach 0.97. The authors highly recommend cost analysis of designed filters KVAR with harmonic and other benefits, periodic system studies especially when new equipments are added to the plant. Also power quality measurements will be necessary to double check harmonics order found through simulation.

A series of tests personal computers in some buildings at King Fahd University of Petroleum and Minerals have been investigated in order to study the influence of these computers on the line current harmonics. The following conclusions can be drawn from the results of this study.

The switch mode power supply (SMPS) used in personal computers draws a non linear current that is rush in harmonics currents. A high density of (SMPS) loads results in over loading of the neutral conductor and the overheating of the distribution transformers.

The assessment of odd harmonics in current significant in magnitudes are represented by mathematical modeling a proved theoretically the decrease in THD in current at some points when increasing the number of PC's connected to these points. On the other hand, THD increased with increase the number of PC's on the other points of these buildings. According to this study the maximum THD found was 29% in the main student lab in building 14 and it was unstable and the minimum THD was found 1.1% in building 58.

According to the instructions provided with the power quality analyzer Fluke 43 B manual which state that if the current THD is less than 20% the harmonic distortion is probably acceptable, the total harmonic distortion at point three of building 14 (29%) is greater than 20% is not acceptable and makes affect on the neutral line cable. To avoid the injection of harmonics into the system, a harmonic filter must be installed.

Due to the highly non sinusoidal nature of the input current waveform of personal computer, the high amplitude of harmonics currents are generated. These harmonics currents are of odd order because of half wave symmetry of the input current waveform. The magnitudes of the harmonics currents up to the seventh harmonics are significant.

The phase angle of the harmonics currents of the input currents of different PC's vary to cause significant current cancelation. There are some cancelations in the higher order harmonics.

The UPS (Uninterruptable power supply) in building 14 can be considered to fit 'in-line' between the loads and the mains power supply. In addition, to providing power protection to the loads, it should also protect the mains power supply itself from any harmonics generated by the loads themselves. However, it is again not commonly known that UPS themselves, by the way of their design, also generate harmonic pollution. For any UPS this is typically stated as Total Harmonic Distortion (THD). The care are has to be taken when comparing different THD values as these can differ when contrasting the two different types of on-line UPS (transformer-based and transformer less) and also with regard to the percentage of load applied for each measurement.

Within a UPS it is the rectifier that connects to the mains power supply and converts the mains alternating current (ac) into the levels of direct current (dc) required to power the inverter and charge the battery.

For transformer-based UPS, rectifiers are typically six or twelve-pulse, dependent upon the thyristor number and configuration. A six-pulse rectifier at full load will typically generate a THD of around 29% and a 12-pulse around 8%. To reduce these values further a passive harmonic filter can be installed alongside the UPS. The obvious disadvantages of this approach being increased capital cost, wiring, installation, loss of efficiency and increased footprint. Harmonic filters can be added post-installation but further installation costs and downtime need to be planned for.

According to the above results obtained from this study, THD at point 2 (29 %) of building 14 does not guarantee with IEEE 519 standers (< 20%) this well cause to reduce the life time of the transformers and cables in building 14 .

## 12. References

[1] J. Arrillaga, D. A. Bradley, and P. S. Bodger, "Power System Harmonics", John Wiley & Sons, New York, 1985.

[2] "Recommended Practices and Requirements for Harmonic Control in Electric Power Systems", IEEE Standard 519-1992, IEEE, New York, 1993.

[3] G. T. Heydt, Electric Power Quality, Stars in Circle Publications, West LaFayette, IN, 1991.

[4] R. C. Dugan, "Simulation of Arc Furnace power systems", IEEE Trans. on Industry Application, Nov/Dec 1980, pp. 813-818.

[5] M. F. McGranaghan, R. C. Dugan, and H. W. Beaty. "Electrical Power Systems Quality", New York: McGraw-Hill, 1996.

[6] Task force on Harmonics Modeling and Simulation, "The modeling and simulation of propagation of harmonics in electric power networks Part I: Concepts, models and simulation techniques," IEEE Transactions on power Delivery, Vol. 11, NO.1, January 1996, pp.

[7] Victor A. Ramos JR, "Treating Harmonics In Electrical Distribution system" Technical Consultant Computer Power &Consulting Corporation, January 25, 1999.

[8] M. H. Shwehdi, et Al," Power Factor Essential and causations," IEEE-PES summer Meeting Singapore, July, 2000

[9] Klaus Timm, Hamburg, basic Principals of electric furnaces, Edited by E. Plockinger and O. Etterich, John Wiley and Sons, Ltd, 1985, pp 127- 160.

[10] Hirofumi Akagi, "New Trends in Active Filters for Power Conditioning", IEEE Trans. on Industry Application, Nov/Dec 1996, pp. 1312-1322.

[11] Antonio Silva, "Steel Plant Performance, Power Supply System Design and Power Quality Aspects", 54th Electric Furnace Conference - Dec. 96.

[12] Joseph S. Subjak, Jr. and John S. Mcquilkin, "Harmonics-Causes, Effects, Measurements and Analysis- Update" IEEE Transactions on industry applications, vol. 3, 1989, pp 55-66.

[13] W.R.A, Ryckaert, J.A.L Ghijselen, J.J.M Desmet, J.A.A. Melkebeek, J. Driesen"The influence on Harmonic Propagation of a resistive shunt harmonic impedance location along a distribution feeder and the influence of distributed capacitors", ICHPQ2004 Lake Placid, NewYork.

[14] H. Akagi, "New trends in active filters for power conditioning," IEEE Trans. Ind. Appl., Vol. 32, No. 6, pp. 1312~1322, Nov./Dec. 1996.

[15] A. Esfandiari, M. Parniani, and H. Mokhtari, "A new control strategy of shunt active filters for power quality improvement of highly and randomly varying loads," in Proc. ISIE2004, pp. 1297~1302, France, 2004.

[16] IEEE recommended practices and requirements for harmonic control of electrical power systems, IEEE Std. 519-1992, 1993

[17] G. W. Allen and D. Segall, "Monitoring of computer installation for power line disturbances," *IEEE PES Winter Meeting Conference*, New York, Jan. 1974, Paper C74199-6.

[18] G. W. Allen, "Design of power-line monitoring equipment," *IEEE Trans. Power App. Syst.*, vol. PAS-90, no. 6, Nov./Dec. 1971.

[19] T. S. Key, "Diagnosing power quality-related computer problems", *IEEE Trans. On Industry Applications*, vol IA-15, no.4, July-August 1979, pp381-393.

[20] M. Goldstein and P. D. Speranza, "The quality of U.S. commercial ac power," in *Proc. INTELEC Conf.*, 1982.

[21] R. Odenberg and B. Braskich, "Measurements of voltage and current surges on the ac power line in computer and industrial environments," *IEEE Trans. Power App. Syst.*, vol. PAS-104, no. 10, Oct 1985, pp 2681- 2688.

[22] L. I. Eguiluz, M. Mañana and J. C. Lavandero, "Voltage distortion influence on current signatures in non-linear loads", *Proc of IEEE PES Winter Meeting 2000*, CDROM 0-7803-6423-6.

[23] A. Mansoor, W. M. Grady, R. S. Thallam, M. T. Doyle, S. D. Krein, M. J. Samotyj, "Effect of supply voltage harmonics on the input current ofsingle-phase diode bridge rectifier loads," *IEEE Trans. Power Delivery*, vol. 10, no. 3, July 1995.

[24] D. O. Koval, C. Carter, "Power quality characteristics of computer loads", *IEEE Trans.Industry Applications*, vol. 33, issue 3, May-June 1997, pp. 613-621.

[25] A. Mansoor, W. M. Grady, A. H. Chowdury and M. J. Samotyj, "An investigation of harmonics attenuation and diversity among distributed single-phase power electronic loads", IEEE Trans. Power Delivery, vol. 10, no. 1, January 1995, pp. 467-473.

[26] A. Mansoor, W. M. Grady, P. T. Staats, R. S. Thallam, M. T. Doyle and M. J. Samotyj, "Predicting the net harmonic currents produced by large numbers of distributed single-phase computer loads," IEEE Trans. Power Delivery, vol. 10, no. 4, Oct 1995, pp. 2001-2006.

[27] Capasso, R. Lamedica, A. Prudenzi, "Experimental characterization of personal computers harmonic impact on power quality," Computer Standards & Interfaces 21 (1999), pp. 321-333.

[28] David Chapman, "Power Quality Application Guide: Harmonics Causes and Effects", Copper Development Association, March 2001.

[29] Philip J. Moore and I. E. Portugues, "The Influence of Personal Computer Processing Modes on Line Current Harmonics", *IEEE Transactions on Power Delivery*, Volume: 18, Issue: 4, pp: 1363- 1368, Oct. 2003.

[30] H. O. Aintablian, H. W. Hill, Jr "Harmonic Currents Generated by Personal Computers and their Effects on the Distribution System Neutral Current", *IEEE Industry Applications Society Annual Meeting*, 1993, Canada, Volume: 2, pp: 1483-1489, 2-8 Oct 1993.

[31] Rana Abdul Jabbar Khan and Muhammad Akmal, "Mathematical Modeling of Current Harmonics Caused by Personal Computers", *International Journal of Electrical and Electronics Engineering*, pp: 103-107, 3:2, 2008.

[32] Rana Abdul Jabbar Khan, "Power Quality and On-line Harmonics Monitoring in Power Systems", *PhD thesis*, RMIT University, 2003.

[33] Juan C. Meza, Abdul H. Samra, "A New Technique to Reduce Line-Current Harmonics Generated by a Three-phase Bridge Rectifier", *IEEE Proceedings of Southeastcon '98*, pp: 354-359, 24-26 April 1998.

[34] Serge B. G. Trochain "Compensation of harmonic currents generated by Computers utilizing an innovative active harmonic conditioner", *MGE UPS Systems*, 2000.

# Permissions

The contributors of this book come from diverse backgrounds, making this book a truly international effort. This book will bring forth new frontiers with its revolutionizing research information and detailed analysis of the nascent developments around the world.

We would like to thank Dr. Gregorio Romero Rey and Dra. Mª Luisa Martinez Muneta, for lending their expertise to make the book truly unique. They have played a crucial role in the development of this book. Without their invaluable contribution this book wouldn't have been possible. They have made vital efforts to compile up to date information on the varied aspects of this subject to make this book a valuable addition to the collection of many professionals and students.

This book was conceptualized with the vision of imparting up-to-date information and advanced data in this field. To ensure the same, a matchless editorial board was set up. Every individual on the board went through rigorous rounds of assessment to prove their worth. After which they invested a large part of their time researching and compiling the most relevant data for our readers. Conferences and sessions were held from time to time between the editorial board and the contributing authors to present the data in the most comprehensible form. The editorial team has worked tirelessly to provide valuable and valid information to help people across the globe.

Every chapter published in this book has been scrutinized by our experts. Their significance has been extensively debated. The topics covered herein carry significant findings which will fuel the growth of the discipline. They may even be implemented as practical applications or may be referred to as a beginning point for another development. Chapters in this book were first published by InTech; hereby published with permission under the Creative Commons Attribution License or equivalent.

The editorial board has been involved in producing this book since its inception. They have spent rigorous hours researching and exploring the diverse topics which have resulted in the successful publishing of this book. They have passed on their knowledge of decades through this book. To expedite this challenging task, the publisher supported the team at every step. A small team of assistant editors was also appointed to further simplify the editing procedure and attain best results for the readers.

Our editorial team has been hand-picked from every corner of the world. Their multi-ethnicity adds dynamic inputs to the discussions which result in innovative outcomes. These outcomes are then further discussed with the researchers and contributors who give their valuable feedback and opinion regarding the same. The feedback is then collaborated with the researches and they are edited in a comprehensive manner to aid the understanding of the subject.

Apart from the editorial board, the designing team has also invested a significant amount of their time in understanding the subject and creating the most relevant covers. They scrutinized every image to scout for the most suitable representation of the subject and create an appropriate cover for the book.

The publishing team has been involved in this book since its early stages. They were actively engaged in every process, be it collecting the data, connecting with the contributors or procuring relevant information. The team has been an ardent support to the editorial, designing and production team. Their endless efforts to recruit the best for this project, has resulted in the accomplishment of this book. They are a veteran in the field of academics and their pool of knowledge is as vast as their experience in printing. Their expertise and guidance has proved useful at every step. Their uncompromising quality standards have made this book an exceptional effort. Their encouragement from time to time has been an inspiration for everyone.

The publisher and the editorial board hope that this book will prove to be a valuable piece of knowledge for researchers, students, practitioners and scholars across the globe.

# List of Contributors

**Soliman Abdelhady Soliman**
Misr University for Science and Technology, Egypt

**Ahmad Mohammad Alkandari**
College of Technological Studies, Kuwait

**Jarosław Łuszcz**
Gdańsk University of Technology, Poland

**Alexandre Brandao Nassif**
Hydro One Inc., Toronto, ON, Canada

**Adrian Plesca**
Gheorghe Asachi Technical University of Iasi, Romania

**Nikolay Bankov, Aleksandar Vuchev and Georgi Terziyski**
University of Food Technologies – Plovdiv, Bulgaria

**Qing Fu, Guilong Ma and Shuhua Chen**
Sun Yat-sen University, China

**Mohsen Abbas Pour Seyyedi and Amir Hossein Jahanikia**
Mefragh Company, Iran

**Luis Sainz and Sara Riera**
Department of Electrical Engineering, ETSEIB-UPC, Spain

**Eduardo Caro**
Department of Electrical Engineering, GSEE-UCLM, Spain

**Miguel García-Gracia, Nabil El Halabi, Adrián Alonso and M.Paz Comech**
CIRCE (Centre of Research for Energy Resources and Consumption), University of Zaragoza, Spain

**Angela Iagăr**
Politechnica University Timişoara, Romania

**M.H. Shwehdi**
King Faisal University, College of Engineering, Al Ahsa, Saudi Arabia

Printed in the USA
CPSIA information can be obtained
at www.ICGtesting.com
JSHW011455221024
72173JS00005B/1082

9 781632 403483